MIND
PERFORMANCE
HACKS™

Other resources from O'Reilly

"*Mind Hacks* is a rewarding mind trip, one that stretches the boundaries of what we know about how the brain works and playfully presents that information in an engaging, thought-provoking way."

—*Blogcritics.org*

Hacks Series Home   *hacks.oreilly.com* is a community site for developers and power users of all stripes. Readers learn from each other as they share their favorite tips and tools for Mac OS X, Linux, Google, Windows XP, and more.

oreilly.com   *oreilly.com* is more than a complete catalog of O'Reilly books. You'll also find links to news, events, articles, weblogs, sample chapters, and code examples.

oreillynet.com is the essential portal for developers interested in open and emerging technologies, including new platforms, programming languages, and operating systems.

Conferences   O'Reilly brings diverse innovators together to nurture the ideas that spark revolutionary industries. We specialize in documenting the latest tools and systems, translating the innovator's knowledge into useful skills for those in the trenches. Visit *conferences.oreilly.com* for our upcoming events.

Safari Bookshelf (*safari.oreilly.com*) is the premier online reference library for programmers and IT professionals. Conduct searches across more than 1,000 books. Subscribers can zero in on answers to time-critical questions in a matter of seconds. Read the books on your Bookshelf from cover to cover or simply flip to the page you need. Try it today for free.

MIND
PERFORMANCE
HACKS™

Ron Hale-Evans

O'REILLY®

Beijing · Cambridge · Farnham · Köln · Paris · Sebastopol · Taipei · Tokyo

Mind Performance Hacks™

by Ron Hale-Evans

Copyright © 2006 O'Reilly Media, Inc. All rights reserved.
Printed in the United States of America.

Published by O'Reilly Media, Inc., 1005 Gravenstein Highway North,
Sebastopol, CA 95472.

O'Reilly books may be purchased for educational, business, or sales promotional use. Online editions are also available for most titles (*safari.oreilly.com*). For more information, contact our corporate/institutional sales department: (800) 998-9938 or *corporate@oreilly.com*.

Editor: Brian Sawyer
Production Editor: Reba Libby
Copyeditor: Audrey Doyle
Proofreader: Sanders Kleinfeld

Indexer: Lucie Haskins
Cover Designer: Linda Palo
Interior Designer: David Futato
Illustrators: Robert Romano, Jessamyn Read, and Lesley Borash

Printing History:

February 2006:   First Edition.

 This book uses RepKover™, a durable and flexible lay-flat binding.

ISBN-10: 0-596-10153-8
ISBN-13: 978-0-596-10153-4
[M]   [11/06]

Contents

Credits

About the Author

Ron Hale-Evans is a writer, thinker, and game designer who earns his daily sandwich with frequent gigs as a technical writer. He has a bachelor's degree in psychology from Yale, with a minor in philosophy. Thinking a lot about thinking led him to create the Mentat Wiki (*http://www.ludism.org/mentat*), which led to this book. You can find his multinefarious [sic] other projects at his home page (*http://ron.ludism.org*), including his award-winning board games, a list of his Short-Duration Personal Saviors, and his blog. Ron's next book will probably be about game systems, especially since his series of articles on that topic for the dear, departed *Games Journal* (*http://www. thegamesjournal.com*) has been relatively successful among both gamers and academics. If you want to email Ron the names of some gullible publishers, or you just want to bug him, you can reach him at *rwhe@ludism.org* (rhymes with *nudism* and has nothing to do with *Luddism*).

About the Developmental Editor

Marty Hale-Evans lives halfway between Seattle and Tacoma, in a perpetually untidy apartment with Mr. Big-Shot Author and two basically useless (but adorable) old dogs. Her professional title is usually "technical editor," under which she has done freelance and contract work for companies such as Microsoft, Boeing, WGBH Educational Foundation, and the University of Chicago Press. Between gigs, she spends her time designing and making jewelry; studying art, history, and Japanese; and reading and writing a lot of email (to and from *marty@martynet.org*). She gets passionate about fat lib, feminism, and politics in general. She has an insatiable comedy jones and is usually reading about four books simultaneously. You can also find Marty playing a lot of board games, sharpening her Texas Hold 'Em skills, singing along, joking around, idly fantasizing about organizing her life, and Trying To Figure It All Out.

Contributors

The following people contributed their hacks, writing, and inspiration to this book:

- Vaughan Bell has just completed a doctoral course in neuropsychology and is now training as a clinical psychologist. When not trying to figure out what goes wrong with the mind and brain, he writes about human behavior and mental life for magazines, books, journals, and the Web.

- Richard Brzustowicz has worked (in no particular order) as a psychotherapist, consulting librarian (information broker), contract writer and editor, emergency room social worker, teacher of English (in Taiwan and Japan), and minor (hopefully not petty) bureaucrat in the world of research administration. His long-standing interest in the byways of psychology led to his research and writing for George Csicsery's film *Hungry for Monsters* (*http://www.zalafilms.com/films/hungryformonsters.html*), about a family caught up in a storm of ritual abuse accusations.

- James Crook is a software engineer who has worked on satellite operating systems, drug discovery, electrical network monitoring, and inside, above, and below TCP/IP stacks. He has a long-standing and deep interest in the mind.

- Karl Erickson is a writer.

- Meredith Hale is the education program manager at the Museum of Glass: International Center for Contemporary Art in Tacoma, Washington. When she's not training docents, writing curricula, and teaching folks about glass and contemporary art, she's working on her master's degree in library and information science. Before committing herself to education, she was a local NPR reporter and a TV news producer, researching and writing stories for air. Although she has written many curricula, news broadcasts, articles for museum publications, and children's book reviews, this is her first time contributing to a real-live book.

- Lion Kimbro is famous for writing "How to Make a Complete Map of Every Thought You Think." He is a Free Software and Free Culture activist, thinker, and programmer. He currently works on wiki and collaboration software.

- Moses Klein's principal passion has been mathematics for as long as he can remember. In 1985, he won a bronze medal at the International Mathematics Olympiad (IMO). More recently, he has graded papers for the IMO, taught mathematics at three colleges and universities, and translated two advanced math texts from French into English.

- Mark Purtill has degrees in mathematics from Caltech (BS, 1984) and MIT (PhD, 1990) and taught at the college level for several years. Currently, he is a senior developer at The Software Revolution, Inc. In his spare time, he plays games and draws pictures of pigs (*http://pigsand toasters.comicgenesis.com*).

- Mark Schnitzius has a degree in computer science, and he has been writing software ever since his father brought home a KIM-1 in 1977. Since then, he has worked at the Kennedy Space Center, the Pentagon, and an on-demand book printing business. He is a six-time winner of the International Obfuscated C Code Competition (*http://www.ioccc.org*) and currently resides in Melbourne, Australia, with his wife and dingo.

- Tom Stafford likes finding things out and writing things down. Several years of doing this in the Department of Psychology at the University of Sheffield resulted in a PhD. Now sometimes he tells people he's a computational cognitive neuroscientist and then talks excitedly about neural networks. Lately he's begun talking excitedly about social networks, too. As well as doing academic research, he has worked as a freelancer, writing and working at the BBC as a documentary researcher. With Matt Webb, he is the author of O'Reilly's *Mind Hacks* (*http://www. mindhacks.com*), a book about do-it-at-home demonstrations of how your brain works. He puts things he finds interesting on his web site at *http://www.idiolect.org.uk*.

- Matt Webb engineers, designs, and works with technology and physical things at Schulze & Webb (*http://www.schulzeandwebb.com*), for clients and for fun. He is coauthor of O'Reilly's *Mind Hacks* (*http://www. mindhacks.com*), a successful cognitive psychology book for a general audience. In the past, he has worked in R&D on social welfare at BBC Radio & Music Interactive on social software, built collaborative online toys, written IM bots, and run a fiction web site (archived at *http://iam. upsideclown.com*). He keeps his weblog, Interconnected, at *http:// interconnected.org/home*. Matt reads a little too much, likes the word *cyberspace*, lives in London, and tells his mother he's "in computers."

Acknowledgments

First, thanks to the authors of *Mind Hacks*, Tom Stafford and Matt Webb, for laying the road and showing the way, and to all of those mnemonists and mental mathematicians out there, ditto.

Next, a hearty thanks to all of the contributors to this book. Man, you guys work hard! Particular thanks to Richard Brzustowicz, lead technical reviewer, and the Marks—Mark Purtill and Mark Schnitzius—who contributed many of the math hacks and tech reviewed the others.

The Mentat Wiki contributors must be mentioned. Thanks, folks, for all of your hard work. You were an inspiration. I would especially like to thank the mysterious StylusEpix for his contribution of the basis for "Count to a Million on Your Fingers" [Hack #40]. He has no known email address and never answered the Wiki messages I left, but maybe he'll see this.

Thanks to Merlin Mann of 43 Folders and 5ives for reviewing the Mentat Wiki, pointing O'Reilly in my direction, and making me laugh so damn hard. Thanks also to Lion Kimbro for encouraging me to create the Mentat Wiki in the first place.

Thanks to my friends for putting up with my neglect while I was writing this book, and especially my board game group, Seattle Cosmic, for putting up with my many absences and foisting unfinished material on them at unforeseen moments. Thanks especially to John Braley for a chess example for the book, and for reminding me to keep my nose clean.

Of the O'Reilly team, thanks to Rael Dornfest and Lucas Carlson for the software I used to write this book, even though I grumbled about it. And a big thanks to my editor, Brian Sawyer. Brian, what can I say? Thank you for your editing, thank you for your gentle introduction to the O'Reilly way, thank you for sorting out rights issues early on, and thank you for your patience as I missed multiple deadlines. (Notice I didn't write anything about thinking *fast*.)

Thanks to my family for encouragement: Mom, Dad, Pam, Eric, Tia, Gwenyth, Mer, Kisa, Mel, and Keith, and especially Darlene for unending positive vibes.

Last but first, this book would still be half a page of scribbled lines without my adorable wife and editor, Marty Hale-Evans, Jeeves to my Wooster, Miss Plimsoll to my Matthew Griswald, fount of gumption, expunger of doofusosity, guardian of good sense, beloved companion, and best friend. I mean, you have *no* idea.

Preface

Think of this book as a martial arts course for your mind. You might call it the *mental arts*. Just as the martial arts will teach you to fight without weapons, the mental arts will teach you to think without computers, calculate without calculators, and remember without reading and writing. Of course, some martial arts courses teach you to fight with nunchaku or throwing blades; this book contains a few hacks that will teach you to use notebooks and Perl scripts as tools to become a better thinker.

In many respects, *Mind Performance Hacks* is a sequel to *Mind Hacks* by Tom Stafford and Matt Webb, but written with the intention of providing practical hacks you can *use*, rather than "probes into the operation of the brain" that are more theoretical and *fun*. (Not that the hacks in this book aren't fun!) Here again, the martial arts analogy comes in handy. *Mind Hacks* (also published by O'Reilly) was more a book of beautiful and interesting things you can do with your brain, akin to the branches of the martial arts that are more like dance than combat. *Mind Performance Hacks* is full of highly practical mental techniques, more analogous to the martial arts meant for self-defense. Of course, this book does contain a few tricks you can flaunt, such as counting to a million on your fingers **[Hack #40]**—the mental equivalent of breaking a stack of bricks with your hands.

In my effort to become a better thinker, besides the raw impulse to become smarter, I think Frank Herbert's idea of the *mentat* has inspired me most. In Herbert's science fiction novel *Dune*, set thousands of years in the future, computers have been outlawed and a human profession has evolved to take their place. Mentats are, in effect, human computers, having trained for years to improve their memory, and their mathematical, logical, and strategic abilities, to superhuman levels—in short, to become masters of the mental arts.

In 2004, I started the Mentat Wiki (*http://www.ludism.org/mentat*) as a central repository of information about the mental arts that anyone who happened upon could contribute to. As I had hoped, it attracted a group of contributors interested in mnemonic techniques, mental math, and so on, and since then, the wiki has grown steadily. I've learned a lot from the Mentat Wiki, and a number of the mind performance hacks in this book started as web pages there.

I hope you will find at least one hack from this book that you use every day for the rest of your life. If that seems unlikely to you, let me assure you that I use several hacks from this book daily, and I know other people who do, too. Have we attained the superhuman capabilities of Frank Herbert's fictional mentats? No, but we are more capable than we would be otherwise.

It's also a dream of mine that the mental arts will be taught commonly in public schools, as the *ars memorativa* was once taught in classical times. Failing that, it would at least be nice if this knowledge were available in little storefront schools emblazoned with the words "Eight Mental Arts Taught As One!"

Why Mind Performance Hacks?

The term *hacker* has a bad reputation in the press. They use it to refer to someone who breaks into systems or wreaks havoc with computers as their weapon. Among people who write code, though, the term *hack* refers to a "quick-and-dirty" solution to a problem, or a clever way to get something done. And the term *hacker* is taken very much as a compliment, referring to someone as being *creative*, having the technical chops to get things done. The Hacks series is an attempt to reclaim the word, document the good ways people are hacking, and pass the hacker ethic of creative participation on to the uninitiated. Seeing how others approach systems and problems is often the quickest way to learn about a new technology.

Mind performance hacks are a technology as new as the newest smart drugs and as old as language. In the broadest sense, every time you learn something, you're hacking your brain. This book is designed to help you learn to hack your brain intentionally, safely, and productively.

How to Use This Book

You can read this book from cover to cover if you like, but each hack stands on its own, so feel free to browse and jump to the different sections that interest you most. If there's a prerequisite that you need to know about, a cross-reference will guide you to the right hack.

You can attempt most of the hacks in this book with no more than your brain, or at least no more than pen and paper. Many contain references to additional material available in other books or on the Web.

A few hacks also include source code for short computer programs—mainly Perl scripts—that you can use as thinking tools. Don't worry, though; you don't need any programming experience to use these hacks. The "How to Run the Programming Hacks" section later in this preface will get you started, and the hacks themselves explain in detail how to install and run the code.

How This Book Is Organized

The book is divided into several chapters, organized by subject:

Chapter 1, *Memory*
> This chapter examines ways to improve your ability to remember information ranging from the periodic table of elements to phone numbers, maps, and spatial locations—as well as how not to leave your keys and your cell phone at home.

Chapter 2, *Information Processing*
> It has been remarked that we live not in an *information* economy but in an *attention* economy. If you have a broadband Internet connection, for example, you have access to more information than you could ever use, but your *attention* is comparatively scarce. This chapter contains hacks to minimize the demands on your attention by maximizing your ability to process this deluge of information.

Chapter 3, *Creativity*
> Creative thought means the generation of new ideas, from making the space shuttle safe to writing a poem for your kid's birthday. This chapter contains some useful techniques for almost any creative challenge you might encounter.

Chapter 4, *Math*
> This chapter examines basic hacks to do mental math, including the *four-banger* operations (addition, subtraction, multiplication, and division), how to check your math, how to count to large numbers on your fingers, and some practical applications, such as calculating the day of the week for any day on the calendar.

Chapter 5, *Decision Making*
> Making decisions with limited data is a problem that everyone faces. The hacks in this chapter will help you separate high-priority from low-priority issues and decide how to take action on those priorities.

Chapter 6, *Communication*

So, you've got some great ideas. How are you going to get them across? This chapter will teach you to do so in ways that are clear, creative, or cryptic.

Chapter 7, *Clarity*

Passionate emotions can cloud our intellectual clarity, causing us to think poorly. Our thoughts can also be riddled with fallacies and self-contradictions. This chapter contains hacks for gaining the emotional and intellectual clarity and perspective needed to solve problems and make good decisions.

Chapter 8, *Mental Fitness*

This chapter contains hacks intended to keep your brain strong and flexible overall, no matter what your age.

Conventions Used in This Book

The following is a list of the typographical conventions used in this book:

Italics

Used to indicate URLs, filenames, filename extensions, and directory/folder names. For example, a path in the filesystem will appear as */Developer/Applications*.

Constant width

Used to show code examples, equations, logarithms, the contents of files, and console output, as well as the names of variables, commands, and other code excerpts.

Constant width bold

Used to show user input in and to highlight portions of code, typically new additions to old code.

Constant width italic

Used in code examples and tables to show sample text to be replaced with your own values.

Gray type

Used to indicate a cross-reference within the text.

You should pay special attention to notes set apart from the text with the following icons:

This is a tip, suggestion, or general note. It contains useful supplementary information about the topic at hand.

This is a warning or note of caution, often indicating that your money or your privacy might be at risk.

The thermometer icons, found next to each hack, indicate the relative complexity of the hack:

beginner   moderate   expert

How to Run the Programming Hacks

The few programmatic hacks in this book run on the command line (that's the Terminal for Mac OS X folks, and the DOS command window for Windows users). Running a hack on the command line invariably involves the following steps:

1. Type the program into a garden variety text editor: Notepad on Windows, TextEdit on Mac OS X, vi or Emacs on Unix/Linux, or anything else of the sort. Save the file as directed—usually as *scriptname.pl* (the *pl* bit stands for Perl, the predominant programming language used in *Mind Performance Hacks*).

   Alternately, you can download the code for all of the hacks online at *http://www.oreilly.com/catalog/mindperfhks*. There you'll find a zip archive filled with individual scripts already saved as text files.

2. Get to the command line on your computer or remote server. In Mac OS X, launch the Terminal (Applications → Utilities → Terminal). In Windows, click the Start button, select Run..., type command, and hit the Enter/Return key on your keyboard. In Unix...well, we'll just assume you know how to get to the command line.

3. Navigate to where you saved the script at hand. This varies from operating system to operating system, but usually involves something like cd ~/Desktop (that's your Desktop on the Mac).

4. Invoke the script by running the programming language's interpreter (e.g., Perl) and feeding it the script (e.g., *scriptname.pl*), like so:

   ```
   $ perl scriptname.pl
   ```

   Most often, you'll also need to pass along some parameters—your search query, the number of results you'd like, and so forth. Simply drop them in after the script name, enclosing them in quotes if they're more than one word or if they include an odd character or three:

   ```
   $ perl scriptname.pl '"much ado about nothing" script' 10
   ```

The results of your script are almost always sent straight back to the command-line window in which you're working, like so:

```
$ perl scriptname.pl '"much ado about nothing" script' 10
    1. "Amazon.com: Books: Much Ado About Nothing: Screenplay ..."
    [http://www.amazon.com/exec/obidos/tg/detail/-/0393311112?v=glance]
    2. "Much Ado About Nothing Script" [http://www.signal42.com/much_
    ado_about_nothing_script.asp]
    ...
```

> The ellipsis points (...) signify that we've cut off the output for brevity's sake.

To prevent the output from scrolling off your screen faster than you can read it, on most systems you can *pipe* (redirect) the output to a little program called *more*:

```
$ perl scriptname.pl | more
```

Hit ^ and the Enter/Return key on your keyboard to scroll through line by line, the spacebar to leap through page by page.

You'll also sometimes want to direct output to a file for safekeeping, importing into your spreadsheet application, or displaying on your web site. This is as easy as:

```
$ perl scriptname.pl > output_filename.txt
```

And to pour some input into your script from a file, simply do the opposite:

```
$ perl scriptname.pl < input_filename.txt
```

Don't worry if you can't remember all of this; each programmatic hack has a "Running the Hack" section that shows you just how it's done.

> Fancy trying your hand at a spot of programming? O'Reilly's best-selling *Learning Perl* (*http://www.oreilly.com/catalog/learnperl4*) by Randal L. Schwartz, Tom Phoenix, and brian d. foy provides a good start.

Using Code Examples

This book is here to help you get your job done. In general, you may use the code in this book in your programs and documentation. You do not need to contact us for permission unless you're reproducing a significant portion of the code. For example, writing a program that uses several chunks of code from this book does not require permission. Selling or distributing a CD-ROM of examples from O'Reilly books *does* require permission. Answering a question by citing this book and quoting example code does not require permission. Incorporating a significant amount of example code from this book into your product's documentation *does* require permission.

We appreciate, but do not require, attribution. An attribution usually includes the title, author, publisher, and ISBN. For example: "*Mind Performance Hacks* by Ron Hale-Evans. Copyright 2006 O'Reilly Media, Inc., ISBN 0-596-10153-8."

If you feel your use of code examples falls outside fair use or the permission given here, feel free to contact us at *permissions@oreilly.com*.

Safari® Enabled

 When you see a Safari® Enabled icon on the cover of your favorite technology book, that means the book is available online through the O'Reilly Network Safari Bookshelf.

Safari offers a solution that's better than e-books. It's a virtual library that lets you easily search thousands of top tech books, cut and paste code samples, download chapters, and find quick answers when you need the most accurate, current information. Try it for free at *http://safari.oreilly.com*.

How to Contact Us

We have tested and verified the information in this book to the best of our ability, but you may find that features have changed (or even that we have made mistakes!). As a reader of this book, you can help us to improve future editions by sending us your feedback. Please let us know about any errors, inaccuracies, bugs, misleading or confusing statements, and typos that you find anywhere in this book.

Please also let us know what we can do to make this book more useful to you. We take your comments seriously and will try to incorporate reasonable suggestions into future editions. You can write to us at:

O'Reilly Media, Inc.
1005 Gravenstein Highway North
Sebastopol, CA 95472
(800) 998-9938 (in the U.S. or Canada)
(707) 829-0515 (international/local)
(707) 829-0104 (fax)

To ask technical questions or to comment on the book, send email to:

bookquestions@oreilly.com

The web site for *Mind Performance Hacks* lists examples, errata, and plans for future editions. You can find this page at:

http://www.oreilly.com/catalog/mindperfhks

For more information about this book and others, see the O'Reilly web site:

http://www.oreilly.com

Got a Hack?

To explore Hacks books online or to contribute a hack for future titles, visit:

http://hacks.oreilly.com

Memory
Hacks 1–12

Memory is a crucial human capability. Without memory, your mind is nothing but bare awareness. Memory orients us in time and space, enables us to recognize our loved ones, provides us with the knowledge that running in front of cars is dangerous, and gives us the raw materials we need to do everything else we do as humans—hence its primary place in this book.

In a sense, many people have abandoned memory, not only to reading and writing, but also to newer technologies such as search engines. However, I hope this chapter will show that developing your memory can enrich your life, whether you need to defend your doctoral thesis, appear on *Jeopardy!*, or just cope with daily hassles.

HACK #1

Remember 10 Things to Bring

You need never forget your keys again. Always remember the top 10 things to bring when you leave your house.

Sure, thanks to the hacks in this chapter on memory, you'll be able to remember all the U.S. presidents and world capitals, but maybe you'll still forget your keys and your cell phone when you leave the house. What good are mnemonic tricks if you can't apply them to daily life?

You can make a practical difference in your preparedness for daily life and the efficiency with which you live it if you memorize a list of items *without which you never leave the house*. If you run through this checklist when leaving work, school, a restaurant, or a friend's house, you need never leave anything important behind wherever you go. You can also use this hack to get out of the house quickly in the morning, by ensuring that all of the items on the checklist are gathered in one place before you go to sleep.

In Action

For this hack, you'll need some kind of mnemonic skeleton that can contain about 10 items (or as many as are on your checklist). You can use a short journey [Hack #3], the 10 digits of the Dominic System [Hack #6], the number shape system [Hack #2], or anything else that *you can remember effortlessly* and when distracted. I use the first mnemonic system I ever learned, the *number rhyme system*, which my father taught me when I was a boy: "One is gun; two is shoe; three is tree," and so on. Ergo, for the first item on my list, I create a vivid image that contains the item and a gun; I remember the second item by associating it with a shoe; and so on down the list.

Compile your checklist and write the items next to the mnemonic skeleton. Put your most important items first in the list so that you'll remember to grab those even if you are interrupted and can't run through your entire list.

As always, link the objects you want to remember to the places in the mnemonic skeleton using the most vivid images you can. Here is my actual list:

1 :: gun :: medication
   I never leave the house without this. I imagine a gun firing pills scatter-shot in all directions.

2 :: shoe :: keys
   I imagine the Old Woman Who Lived in a Shoe trying to open the front door of her giant shoe with her keys while dozens of her children are tugging on her skirt.

3 :: tree :: cell phone
   I imagine a tree with a 1920s-style varnished black telephone handset and mouthpiece protruding from it. A pair of bells on the tree ring loudly.

4 :: door :: notebook
   I imagine my Moleskine Mini notebook grown to enormous size. The front cover swings open like a door with a huge *Inner Sanctum* creak of hinges. (I never go anywhere without my catch [Hack #13].)

5 :: hive :: wallet
   I imagine opening my wallet and a swarm of bees flying out into my face. Argh!

6 :: sticks :: PDA
   I imagine using a scratchy wooden stick with leaves as a stylus to write on my PDA. (This also reminds me to bring a stylus, in case I forget in step 10.)

7 :: heaven :: eyeglasses

> I imagine my eyeglasses shining, because they are made out of the same nacreous material as the Pearly Gates.

8 :: gate :: handkerchief

> I imagine my handkerchief tied to the post of an ordinary garden gate and flapping in the wind like a flag as the gate swings back and forth.

9 :: wine :: Swiss Army knife

> I imagine that one of the blades of my red Swiss Army knife is actually a miniature wine bottle and that when I open it, a flood of red wine pours out.

10 :: hen :: pen

> I imagine that a hen is pecking at a bunch of the four-color pens I use, which lie about on the ground in abundance with some PDA styluses.

In Real Life

I can honestly say that in the year or two I have been using this technique, I have forgotten particular items on the list only once or twice, and that was merely because I didn't run through the whole list as I was packing up. There was, however, one incident, where my wife and I were late because I set my fully packed bag down momentarily and forgot to bring the whole kit with me!

I'm now so familiar with my list that *10* brings *pen* to mind immediately rather than *hen*, and so on. Therefore, it's been easy to add two more items without thinking of additional mnemonics for them: 11, my exoself [Hack #17], and 12, a good book to read. However, if I were going to extend my checklist any more, I would certainly add more mnemonic pegs.

See Also

- A *murse* (a.k.a. *man purse*), such as that which I carry all my gear in, is an essential mini-hack in itself. Of course, women reading this who've been carrying purses for years will think that men like me are a bit slow and wonder what business we have writing a book on mental performance. Still, the Slacker Manager blog has a great intro to the murse concept; see *http://www.slackermanager.com/slacker_manager/2004/11/my_murse.html*.

- It's also useful to stock bags for different purposes, such as school, work, and emergencies. These work as a kind of external memory; you can just grab them and go.

Use the Number-Shape System
#2   Associate numbers with shapes and use the hunting and gathering faculties
of your primitive ancestors to remember 21st-century data.

If you've learned how to remember 10 things to bring when you leave the
house [Hack #1], you've already learned the number-rhyme system: associating
numbers like 1 and 2 with words that rhyme with them, like gun and shoe,
and using those associations as *pegs* on which to hang items you wish to
remember.

The traditional *number-shape system* works in a similar way. Instead of visualizing images whose names rhyme with the names of numbers, however,
you visualize shapes that look like the numerals in question. For example,
the numeral 2 looks like a swan to many people, so you can use the image of
a swan as a mnemonic peg.[1]

In Action

Table 1-1 lists 10 digits, along with some shapes you can use to remember
them. The Shape column illustrates the italic words in the Words column,
to show how the associations arose.

Table 1-1. Corresponding numbers, words, and shapes

Number	Words	Shape
0	Black hole, donut, *tire*	
1	*Candle*, pencil	
2	*Swan*	
3	*Butterfly*, heart	
4	*Sailboat*	
5	*Hook*, pulley	

Table 1-1. Corresponding numbers, words, and shapes (continued)

Number	Words	Shape
6	Golf club, lasso, *pipe*	
7	*Axe*, boomerang, scythe	
8	Hourglass, *snowman*	
9	*Flag*, tadpole	

Feel free to pick and choose, or devise your own shapes. It's most important to be consistent so that when you want to remember what you associated with the number 6, you don't waste time trying to remember whether your mnemonic shape is a pipe, a lasso, a golf club, or something totally different.

How It Works

Like the brains of all animals, the human brain has a lot more experience with concrete shapes than with abstract numbers. For example, the decimal digit 0 was not even discovered until about 300 BC.[2] Our ancestors used their senses to learn more about the world, find food, escape predators, and perform many other essential tasks. These tasks were vital to our survival in an evolutionary sense, so the faculties involved in processing sensory information were well developed, and today our brains still process this kind of information thoroughly and efficiently.

Thus, by turning numbers into concrete shapes and making them even more imaginably vivid with motion, humor, sex, aggression, color, smell, touch, taste, and all the other features of the real world that our primate brains evolved to process, we're in effect wrapping the numbers with mental friction tape so that we can grasp them better.

Of course, it's possible to become intimately familiar with numbers in their own right [Hack #36], and doing so will help you not only to remember them, but also to get better at math.

In Real Life

Here is how you might use the number shape system to remember a list of five items to pick up at an office supply store. As is often the case with image-based memory systems, the more vivid the mental image you can conjure, the easier it will be for you to remember.

1 :: candle :: a spindle of DVD-Rs
  Imagine a silvery DVD being played by a gramophone with an upside-down candle for a needle, dripping hot wax onto the disc.

2 :: swan :: index cards
  Imagine a white swan whose wing feathers are 3×5-inch index cards.

3 :: butterfly :: printer paper
  Imagine a multicolored butterfly getting stuck to a piece of white printer paper and going through your printer. When it emerges in the output tray, it peels away from the paper and flaps into the distance.

4 :: sailboat :: four-color pens
  Imagine a sailboat whose mast is a gigantic four-color ballpoint pen. As the sailboat tacks with the wind, the pen clicks and different colors emerge.

5 :: pulley :: manila envelopes
  Imagine a huge pulley in the warehouse at the back of the office supply store, swinging an enormous bundle of manila envelopes, bound together with manila twine.

I often use the number shape system to make a quick shopping list when I have to grab some groceries on the way home. If you learn this system, you'll probably develop many applications of your own, such as taking notes on the points you need to address when it's your turn to speak in a meeting.

End Notes

1. Mentat Wiki. "Number Shape System." *http://www.ludism.org/mentat/ NumberShapeSystem*.

2. Wikipedia. "0 (number)." *http://en.wikipedia.org/wiki/0_%28number%29*.

See Also

- The number shape system is also useful when remembering numbers with the Dominic System [Hack #6]. For example, you can remember a five-digit number the usual way you remember a four-digit number, but incorporate a shape for the fifth digit into the image.

Make Lots of Little Journeys

HACK #3

Making mental journeys (also known as "memory palaces") is a useful way to remember sequential information. If you have several familiar short journeys handy, you can be ready to remember whatever you need to, at any time. Here's how to start with the layout of your own house or apartment.

Practically every system of mnemonics relies on a series of *pegs* on which to hang information. For example, "Remember 10 Things to Bring" **[Hack #1]** associates the numbers 1 through 10 with rhyming objects (one = gun, two = shoe, three = tree, and so on) and then hangs the things to remember (such as medication, keys, and cell phone) on these mnemonic pegs by putting the peg objects and the things to remember in the same vivid mental picture.

An even older mnemonic technique—perhaps the oldest—uses *places* as memory pegs. By *places*, I mean ordinary, concrete places, such as the rooms of your house or apartment. If you mentally organize these places into a sequence that is the same every time, you will be able to walk through the places in your mind and retrieve the information you have stored there.[1]

The Renaissance practitioners of the ancient *ars memorativa* (art of memory) referred to such journeys as *memory palaces*. Orators in classical times would prepare their speeches by stashing complex images that represented the things they wanted to talk about in the *loci* (places) of a remembered or imagined building, such as a palace. In fact, this practice is said to be the origin of today's expressions "in the first place," "in the second place," and so on.

In Action

When you create your mental images, make the impressions of the objects you want to remember as vivid as possible, to make the ideas you want to remember *stick* to the places of your journey. You can do this in many ways, such as by exaggerating them or using humor, sex, bright colors, motion, or anything else that holds your attention. (The word *impression* comes from yet another classical metaphor depicting memories as the marks left by a stylus on a wax tablet, the yellow legal pad of the day. When you make impressions on the wax tablet of your mind, *press down hard*.[2])

To assemble your first memory journey, use a place you know extremely well; your home is a good example. You can also use the shops along a street where you walk every day or the benches, brooks, and shady trees of your favorite park. Just make sure you can trace your journey from beginning to end in your mind's eye before you try to use it as a mnemonic tool.

After you memorize one list of objects with your journey, you can "wipe the wax clean" and reuse the journey by mentally walking its length and visualizing the places as being empty of the objects you memorized. Blow up the objects with dynamite if you like.

You might want to create multiple journeys of different lengths, so you have one ready for any occasion. Then, if you need a journey longer than any you have memorized, you can link two or more journeys together by starting one where the last ended; imaginary journeys don't need to obey real-world geography.

In Real Life

Here are the first 10 places on my first journey, with typical actions envisioned for each. This journey is based on my real-life apartment. Places 1–5 start on the right side of the apartment (as seen from the start of the journey). After place 5 (the porch), the journey makes a left turn into the living room and doubles back so that places 1–10 make a horseshoe shape.

1. *Bedroom*
   Where I start my day.
2. *Back bathroom*
   My first stop every morning.
3. *Front bathroom*
   Get some clothes out of the dryer.
4. *Computer room*
   Check my email.
5. *Porch*
   Get a breath of fresh air.
6. *Living room*
   Sit down on the couch.
7. *Dining room*
   Have breakfast.
8. *Kitchen*
   Grab coffee to go.
9. *Entry*
   Grab keys for the car.
10. *Outside*
    On to the next journey?

When I was reading the "Famous Forty" Oz books, starting with *The Wonderful Wizard of Oz* and running through the next 39 canonical books in the series, I used pegs 1–40 of the Dominic System [Hack #6] to memorize their titles. Let's use my apartment journey to memorize the titles of Shakespeare's 10 tragedies:

1. *Bedroom: Titus Andronicus*

   A drill sergeant standing on the bed, wearing pants much too small, is cursing Private Ronald Reagan roundly (*Tight-ass and Ronnie-cuss*).

2. *Back bathroom: Romeo and Juliet*

   A teenage boy and girl (guess who) are necking in a bright red Alfa *Romeo* sports car in the bathtub.

3. *Front bathroom: Julius Caesar*

   An ancient Roman man sits in the bathtub sipping an Orange *Julius* drink and eating a *Caesar* salad.

4. *Computer room: Hamlet*

   Piglet (the character from *Winnie-the-Pooh*) is stranded on the top of my computer monitor (Piglet = *Hamlet*).

5. *Porch: Othello*

   Two men in Shakespearean garb are seated and playing the board game *Othello* on the porch table.

6. *Living room: Timon of Athens*

   There is a tiny baseball *team on* the coffee table. They are *of Athens*: they have long white beards and are declaiming from scrolls.

7. *Dining room: King Lear*

   An old man in a crown (a *king*) is seated at the table. He is *leering* at me, elbowing me in the ribs, and winking.

8. *Kitchen: Macbeth*

   A gigantic *Mac* computer in the sink starts with a "bong!" sound and displays a beautiful picture of Queen Elizabeth I (*Beth*).

9. *Entry: Antony and Cleopatra*

   A woman in ancient Egyptian headgear (*Cleopatra*) sits in front of the door, wrinkling her nose and picking *anchovies* (Antony) off her pizza.

10. *Outside: Coriolanus*

    I find some bright green herbs on the ground. Feh! They're *coriander* (cilantro), which I hate. I wash out my mouth with *anise* seed, which tastes like licorice.

Notice that I used *images* to remind me of specific *words* in the titles of the plays. Since I'm already familiar with the play titles, this should be enough to remind me of them. If you don't already have a rough idea of the things you're trying to memorize, you might need to make more detailed images that are less ambiguous, or piggyback another memory technique onto this one.

The images need not have a logical connection with the mental location where you place them, since the places in the journey are essentially arbitrary, just like the pegs in the number rhyme system [Hack #1]. For example, I don't have a Macintosh in my kitchen sink, nor would I ever put one there. Actually, that very fact makes the placement of an imaginary Mac there all the more memorable. Absurdity is one of the many techniques used to make images vivid.

If you want to remember more play titles, simply add more places to the journey. For example, place 11 could be my car and could hold the first Shakespearean comedy, *The Comedy of Errors*. I could open my car door outside and find that the controls on my dashboard are backward and upside down, which makes me laugh.

Imaginary journeys can be extended indefinitely, so after you memorize all of Shakespeare's plays, you can move on to the works of other authors, or anything else you want to remember.

Try this the next time you are shopping in a familiar place, such as the usual place you buy groceries: mentally plot an efficient path through the store as a memory journey, then pick up what you need and go directly to the cashier. If you normally browse and buy a little too much, this technique may suggest a different approach.

End Notes

1. Mentat Wiki. "Memory Palace." *http://www.ludism.org/mentat/Memory Palace*.

2. Mentat Wiki. "Link Quickly." *http://www.ludism.org/mentat/LinkQuickly*.

See Also

- *The Amazing Memory Kit* (Duncan Baird) by Dominic O'Brien is a useful collection of interactive tools for training your memory. Amusingly, it also contains a sample memory journey for remembering Shakespeare's 10 tragedies; I was unaware of the example while writing this hack.

HACK #4 Stash Things in Nooks and Crannies

Systematically place information in the corners and walls of rooms, and expand the capacity of your memory journeys up to tenfold.

"Make Lots of Little Journeys" [Hack #3] explains how to remember information by associating it with places along the way in an imaginary journey. (If you haven't read that hack, please read it now.) But each place on a memory journey contains other places: rooms typically have four walls, four corners, a floor, and a ceiling, for a total of 10 sublocations. In other words, if you have already memorized a journey through a building, you can now make your memory journey hold 10 times as many pieces of information.[1]

In Action

Scott Hagwood, the U.S. Grandmaster of Memory, seems to have invented the nooks-and-crannies hack. He used it to break the world record for color-sequence memorization for the electronic game Simon. The previous record had been 14 sequences, but Scott was able to play an astonishing 31 sequences—all that the machine could offer. To do so, he used a memory journey and mentally stuffed the corners and walls of his places with items representing the colors he was trying to remember, such as a yellow sun or green bouncy balls.

Reconstructing Hagwood's system from his interviews is simple enough.[2,3] Hagwood's map for each room looks something like Figure 1-1.

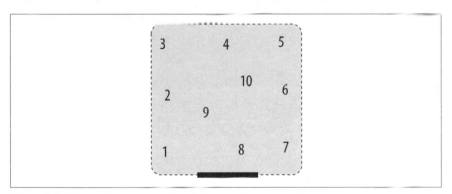

Figure 1-1. The nooks and crannies of Hagwood's memory journey

The map in Figure 1-1 assigns numbers to the following places:

1. Near-left corner
2. Left wall
3. Far-left corner
4. Far wall
5. Far-right corner
6. Right wall
7. Near-right corner
8. Near wall/entrance to room

9. Floor

10. Ceiling

While this might not be Hagwood's exact system, it's the one we'll use in this hack.

In Real Life

In "Make Lots of Little Journeys" [Hack #3], we used a memory journey to recall Shakespeare's tragedies. Suppose you want to store more information about the plays in the same journey. You might associate the features of the plays with the features of the rooms in your memory journey in this way:

1. *Near-left corner*
   Plot event 1.

2. *Left wall*
   Plot event 2.

3. *Far-left corner*
   Plot event 3.

4. *Far wall*
   Plot event 4.

5. *Far-right corner*
   Plot event 5.

6. *Right wall*
   Plot event 6.

7. *Near-right corner*
   Plot event 7.

8. *Near wall/entrance to room*
   Plot event 8 (climax).

9. *Floor*
   Publication date.

10. *Ceiling*
   Related work. (Many later authors have based books or plays on Shakespeare, just as Shakespeare derived his work from earlier authors.)

You can think of the floor and the ceiling as special places within the room, so you can use them for special information about the play—in this case, the publication date and a related work. (The other eight places are all the same: slots for plot elements.) If you were memorizing the periodic table of elements, you might use the floor and ceiling for two key pieces of information: the alphabetic symbol of the element (such as *Au* for gold), and its atomic number (79). The rest of the places could then be used for other details.

Let's try the nooks-and-crannies hack to memorize the details of Shakespeare's second tragedy, *Romeo and Juliet*. In "Make Lots of Little Journeys" [Hack #3], I associated this play with the back bathroom of my apartment. Now I will associate the features of the play with the features of this room, in detail:

1. The Montagues and Capulets brawl (in the dogs' water dish).

2. Paris convinces Juliet's parents to let her marry him (mopping his brow with my towel).

3. Romeo falls in love with Juliet at a masked ball. (Their hair is sudsy with shampoo.)

4. Romeo and Juliet declare their love in the balcony scene and secretly marry (in the bathtub).

5. A duel ensues in the street, in which Tybalt kills Mercutio, and Romeo then kills Tybalt. (The guards wash away the blood with water from the shower.)

6. Juliet quarrels with her father about Paris (down in the toilet bowl).

7. Juliet takes a sleeping potion to feign death. (She falls facedown into the sink.)

8. Climax: Romeo discovers the drugged Juliet in a tomb; he kills Paris, who is mourning her; he poisons himself; she awakes and stabs herself. (All of this occurs in the doorway.)

9. Published in 1594. (Image: Albert Einstein (AE = 15), standing on my bathroom rug, walks into a red sandstone building with a thud (Nick Danger = ND = 94).)

10. The musical *West Side Story*. (Image: a miniature version of the scene from the movie in which the dancers split into many different colors, in the light of the heat lamp.)

The mnemonic for the publication date is worked out using the Dominic System [Hack #6]. The plot events should not be hard to visualize if you've seen the play; otherwise, you can use mnemonic tricks, such as representing Paris with a miniature Eiffel Tower, Mercutio with winged sandals like those of the god Mercury, and so on.

You can use the nooks-and-crannies hack to memorize any information that can be presented serially, from the digits of π, to the telephone area codes of the U.S. and Canada, to the nations of the world in alphabetical order. And if all of these examples leave you unimpressed, does the word *Vegas* suggest anything more to your taste?

End Notes

1. Mentat Wiki. "Nook And Cranny Method." *http://www.ludism.org/mentat/NookAndCrannyMethod.*

2. Zasky, Jason. 2003. "Total Recall: Remembering the 2003 USA Memory Championship." *Failure Magazine*, March 2003. *http://www.failuremag.com/arch_flop_total_recall.html.*

3. Gupta, Sanjay. 2005. "Mystery of Memory." Transcript of CNN report. *http://www.bio.uci.edu/public/press/2005/MysteryMemory.pdf.*

HACK #5   Use the Major System

The Major System is the most commonly used set of mnemonics. This custom Major System will help you memorize lists of up to 100 items, as well as credit card PINs, phone numbers, and the other numeric trivia of daily life.

The Major System was introduced in the 17th century by Stanislaus Mink von Wennsshein and was improved in the 18th century by Dr. Richard Grey.[1] While the Major System is probably the most established mnemonic schema, I prefer the Dominic System [Hack #6], invented by Dominic O'Brien in the 20th century. Nevertheless, you might find that the Major System works well for you, and knowing something about it will contribute to your understanding of advanced mnemonic techniques.

In Action

The Major System uses *peg words* just like the number-rhyme system [Hack #1] and number-shape system [Hack #2]. Instead of associating numbers with peg words based on rhymes or shapes, however, it assigns each digit a basic consonantal sound and builds up peg words from combinations of those consonants. For example, the digit 3 is linked to the consonant M, and the digit 2 is linked to the consonant N, so our Major System list suggests *moon* for 32.

The consonant assignments are fairly arbitrary—Lewis Carroll came up with an alternate set [Hack #9] that's probably just as good—but Table 1-2 shows a standard set of mnemonics you can use for these associations until they become second nature.

Table 1-2. Number/letter associations

Number	Letter	Association
0	S, Z, soft C	*Z* is the first letter of *zero*.
1	D, T, TH	The letters *d* and *t* have only one downward stroke.

Table 1-2. Number/letter associations (continued)

Number	Letter	Association
2	N	The letter *N* has two downward strokes; it also looks like the numeral *2* rotated 90 degrees.
3	M	The letter *M* has three downstrokes; it also looks like the numeral *3* rotated 90 degrees.
4	R	The letter *R* is the last letter in *four*.
5	L	*L* is the Roman numeral for 50; also, a human hand with its thumb stuck out looks like an *L*.
6	J, SH, DG, soft G, CH as in *cheese*	*J* looks like *6* backward.
7	K, hard C, hard G, QU, CH as in *loch*	You can draw a *K* with two *7* characters.
8	F, V	A cursive, lowercase *f* looks like an *8*.
9	B, P	A *b* looks like a *9* rotated 180 degrees; a *P* looks like a backward *9*.

Table 1-3 shows the Major System peg word list for the numbers 1 to 100. If you don't like the words I use, you can use your own. Since your list will use your own mental connections, it might be even more effective for you. Just be consistent, so you don't have to grope to remember the peg word for a particular number.

Table 1-3. Number/word associations

Number	Word	Number	Word
1	Tea, Tie	51	LaD, LiD
2	Noah	52	LioN
3	Ma	53	LamB
4	eaR, heRo, oaR, Rye	54	LuRe
5	Law	55	LiLy
6	Shoe	56	LaSh, LeaSh, LeeCh
7	Key	57	LaKe, LocK, LoG
8	iVy	58	LaVa, LeaF
9	Bee, Pie	59	LiP
10	DiCe, ToeS	60	CheeSe
11	DaD, ToaD, ToT	61	SheeT
12	TiN	62	ChaiN
13	DaM, ToMb	63	ChiMe
14	TiRe	64	ChaiR, CheRry
15	TaiL, ToweL	65	JaiL

Table 1-3. Number/word associations (continued)

Number	Word	Number	Word
16	DiSh	66	Choo-Choo
17	TacK	67	ChalK
18	DoVe, TV	68	CheF
19	TaPe, TuB	69	ShiP
20	NoSe	70	CaSe
21	gNaT, NeT	71	CaT
22	NuN	72	CaN, CoiN
23	eNeMa	73	CoMb
24	NeRo	74	CaR
25	NaiL	75	CoaL
26	hiNGe, NotCh	76	CaGe
27	NaG, NecK	77	CaKe, CoKe
28	kNiFe	78	CaFe, CoFfee, CaVe
29	kNoB	79	CaB, CoB
30	MiCe, MooSe	80	FeZ, VaSe
31	MaiD, MaT	81	VaT
32	MooN	82	FaN, PhoNe
33	MiMe, MuMmy	83	FoaM
34	MoweR	84	FuR
35	MaiL	85	FiLe
36	MatCh	86	FiSh
37	MaC, MuG	87	FoG
38	MaFia, MoVie	88	FiFe
39	MaP, MoP	89	VP (such as Dick Cheney)
40	RiCe, RoSe	90	BuS
41	RaDio, RaT	91	BaT, BoaT
42	RaiN, RhiNo	92	BoNe
43	RaM	93	BoMb
44	RoweR	94	BeaR
45	RaiL, RolI	95	BaLl, BowL
46	RoaCh	96	BeaCh
47	RacK, RaKe, RocK	97	BooK, PiG
48	RooF	98	BeeF
49	RoPe	99	BaBy
50	LaCe, LaSsie	100	DaiSieS

When you're converting numbers to peg words and back, there are some simple rules to follow:

- Ignore vowels.
- Ignore the second consonant in double-consonant sounds. For example, *mummy* is 33, not 333, because the *mm* is counted as a single *m*.
- Ignore silent consonants. For example, *neck* is 27, not 277, because the C does not contribute to the K sound.

 To make these rules more obvious, Table 1-3 capitalizes only the consonants being used to form the peg word. For example, the peg word for 32 is listed as *MooN*.

Since there are only a limited number of possible words in English for each number, the mnemonic lists for the Major System in most memory books tend to be similar. I consulted three of the books in my collection[2,3,4] to compile this list, taking the best from each and adding my own words (such as 42 = RaDio and 81 = VaT) when it seemed useful. I followed several guidelines, which might interest you if you want to customize it:

- I preferred nouns to verbs and adjectives, because they are more easily visualized.
- I preferred concrete objects to abstract objects for the same reason.
- I preferred more active, versatile objects (49 = RoPe, not RuBy), ditto.
- I tried not to interfere with other memory systems in this book (for example, deleting 99 = PiPe because it might interfere with 6 = pipe in the number shape system) or other Major System items (for example, 95 = BeLl interfered with 63 = ChiMe).

In Real Life

If you have a credit card with the number 4880 6630 6767 7584 (these digits were generated at random with the dice-rolling application on my PDA), you might remember them with the words shown in Table 1-4.

Table 1-4. Associations for a credit card number

Number	Word
48	Roof
80	Vase
66	Choo-choo
30	Moose

Table 1-4. *Associations for a credit card number (continued)*

Number	Word
67	Chalk
67	Chalk
75	Coal
84	Fur

But now you have a new problem: how are you going to remember this arbitrary list of words, especially the two sequential instances of *chalk*?

You can use the following little story—a kind of memory journey [Hack #3]—to string them together:

> A shingle falls off your *roof* and breaks a *vase*. A *choo-choo* of the sort in a kiddie amusement park chugs up and the conductor gets out. He's a *moose*! He scolds you and tries to write you up with a piece of *chalk*, but his slate is also made of *chalk*. (How weird.) He bellows in frustration and begins shoveling *coal* to leave, but he gets sooty, and abashedly asks you to brush out his *fur*.

With just a few mental rehearsals of this story (or the equivalent for your own card), you just might never forget your credit card number again. If you actually recall the number a few times in the bustle of real life, you'll begin to remember the number directly and let the story fade naturally.

End Notes

1. Buzan, Tony. 1989. *Use Your Perfect Memory*, Third Edition. Plume.

2. Ibid.

3. Lorayne, Harry. 2000. *How to Develop a Super Power Memory*. Frederick Fell Publishers, Inc.

4. Trudeau, Kevin. 1995. *Kevin Trudeau's Mega Memory*. Quill/William Morrow.

HACK #6   Use the Dominic System

The Dominic System, invented by World Memory Champion Dominic O'Brien, is an easier alternative to the Major System of mnemonics found in most memory books.

Dominic O'Brien, World Memory Champion, can memorize the order of a full deck of playing cards in less than a minute. To help him achieve amazing memory feats like this, he created the Dominic System of mnemonics. Some people who find the Major System [Hack #5] espoused by most memory experts to be too dry and restrictive find they can stick with the Dominic System.

In Action

The Dominic System uses an easy-to-remember number-to-letter conversion and the initials of memorable people, as well as *journeys* that are like memory palaces [Hack #3]. As many mnemonic systems do, the Dominic System requires some bootstrapping for you to reach its full potential.

You will have to spend a little time and work to memorize the structure of the system, and that might seem a little tedious. Your work will be rewarded, however, because this basic work will enable you to harness the system's full power for yourself. It's a little like starting slow on the treadmill at the gym if you want to work up to taking long hikes in the mountains.

The number-to-letter correspondences run as follows:[1]

Digit	Letter
1	A
2	B
3	C
4	D
5	E
6	S
7	G
8	H
9	N
0	O

You can remember the numbers 00 to 99 by linking them to famous people and actions that are characteristic of them. For example, the number 15 becomes AE. You might mentally connect the initials AE with Albert Einstein and assign writing on a blackboard as Einstein's characteristic action. Similarly, 80 = HO = Santa Claus, laughing and holding his belly (HO, HO, HO!). You can use my list[2] or O'Brien's list, but the system will work best if you use the associations that are already in your own mind.[3]

After you have the two-digit associations firmly in your mind, you can remember four-digit numbers by combining the person associated with the first two digits and the action associated with the second two digits. Thus, 8015 can translate to HOAE, which can be broken down to HO and AE. To remember it, think of Santa Claus (HO) with Albert Einstein's action (AE): Santa Claus writing on a blackboard.

You can remember five-digit numbers by adding a symbol from the number-shape system [Hack #2] to the image, so that 80152 might be represented by Santa writing on a blackboard with a swan (2) tucked under one arm. You can remember longer sequences of numbers (such as memorizing the digits of π), or sequences of any kind, by chunking them [Hack #11] and committing them to the places of a memory journey.

Here's an example of how to memorize a 12-digit number: the month table used to calculate weekdays [Hack #43]. I will use my personal associations for the letter combinations and fill in from other sources when my own mnemonics are too idiosyncratic to make sense to most people.

First, make one long list of the month numbers. Since none of them is larger than 6, they are all one digit long, so we obtain a 12-digit number:

    033614625035

Next, break up this list (i.e., *chunk it*) into three four-digit numbers:

    0336 1462 5035

> Four-digit numbers are easy to memorize in the Dominic System. Besides, it's a fairly natural division; for example, all the months that end in *-ber* form the last group.

Now, apply the Dominic System mnemonics:

- 0336 = OCCS = Oliver Cromwell/C.S. Lewis
- 1462 = ADSB = Jesus (AD)/Sandra Bullock in the movie Speed
- 5035 = EOCE = Eeyore/Clint Eastwood

Next, make an imaginary journey [Hack #3] by using the first character associated with each number performing the action associated with the second character:

> Oliver Cromwell (OC) steps into the magic wardrobe and ends up in the land of Narnia (CS). He wanders through the snow until he comes to the lamppost, where he meets Jesus (AD), who leaps into a bus and starts driving away like crazy (SB). Jesus doesn't get far, however, because Eeyore (EO) appears from behind a bush and lassos him like Clint Eastwood (CE).

Did you find this little journey offensive or surreal? Strong emotional reactions help people remember things, so outlandish mental images can actually be more effective to use. Again, you should make your own list of characters for your own version of the Dominic System, and then you can tune your list to suit yourself.

How It Works

The Dominic System is a combination of the innovative (easier mnemonic alphabet, using people rather than inanimate objects because people are easier to remember, etc.) and the tried-and-true (memory palaces, which go back to classical times). It has a couple of advantages over the Major System and its derivatives:

- The 1 = A, 2 = B, 3 = C, and so on, Dominic System is easier to learn than the Major System's more arbitrary 1 = T/D/TH, 2 = N, 3 = M, etc. There is circumstantial evidence that the Dominic System is also faster and more powerful: Dominic O'Brien became World Memory Champion using his system, a title that includes competitions for speed in memorization.
- The famous people of the Dominic System are combined with their characteristic actions in an easy, natural way, making numbers up to four digits long easy to memorize with a single image.
- If you can memorize a four-digit number with the Dominic System, you can memorize 10,000 pieces of information, from 0000 to 9999 [Hack #7].

In Real Life

As my first test of the Dominic System, I used the subset of numbers from 01 to 40 to memorize the titles of a favorite series of books, the so-called Famous Forty by L. Frank Baum and his successors, set in the Marvelous Land of Oz. For example, book 23 is *Jack Pumpkinhead of Oz*. The number 23 corresponds to BC in my personal Dominic System, for which the person/action pair was Thor, the character from the comic strip *B.C.*, riding his stone unicycle. Thus, the image I used to remember this book was Jack Pumpkinhead riding a stone unicycle.

Memorizing the titles of the Famous Forty took about 45 minutes, approximately a title a minute. A minute is about as long as it takes Dominic O'Brien to memorize an entire shuffled deck of cards, so at that point I had a lot of room for improvement. These days, I can memorize items several times faster than I could at the beginning—still not as fast as a World Memory Champion, but improving.

As it happened, I made a fruitful mistake with this test. I was already familiar with the more common Major System, and I thought that the Dominic System's numbers from 00 to 99 were meant to be used as mnemonic *pegs* or places [Hack #3], just like the numbers in the Major System. In fact, the numbers are mainly used to encode numeric information; Dominic himself would probably memorize the Famous Forty with a memory journey.

I'm glad I made my initial mistake, though, because it led me to build on the Dominic System to construct the Hotel Dominic [Hack #7], which theoretically enables you to memorize 10,000 items or more of information.

End Notes

1. O'Brien, Dominic. 1994. *How to Develop a Perfect Memory*. Trafalgar Square. The canonical reference for the Dominic System, and a memory classic. Unfortunately, it is out of print in hardcopy form, and the last copy I spotted (in 2004) cost about $150. Fortunately, it is available less expensively as an e-book here: *http://www.lybrary.com/index.html?goto =books/how_to_develop_memory.html*.

2. A file containing my personal mnemonic pegs is available at *http://ron. ludism.org/mnemonics_public.txt*, including my version of the Dominic System for numbers 00–99. Use this only as an example, since many of the names in the list are idiosyncratic. Some of them refer to friends and family, and I have simply removed them in the public version and replaced them with the word *PERSONAL*.

3. Blank Dominic System template for your own characters. *http://ron. ludism.org/dominic_template.txt*.

See Also

- Matt Vance has created a page (*http://www.minezone.org/wiki/MVance/ DominicSystem*) that lists *multiple* possible characters for each number from 00–99. This is a good place to start when you construct your own list.

- "Visit the Hotel Dominic" [Hack #7].

HACK #7 Visit the Hotel Dominic

Expand the basic Dominic System list of 100 to hold 10,000 items or more of information.

You might need to memorize a table or list with more than 100 elements, such as the periodic table of elements, but find that you can't do it with only the 100 numbered items of the Dominic System [Hack #6]. You could use a memory journey [Hack #3], but how are you going to remember that element 52 is tellurium without visiting the 51 previous rooms first?

This memory hack, which I call the Hotel Dominic (in honor of Dominic O'Brien, the inventor of the Dominic System of mnemonics upon which it's based), is both *random access* (like a CD, as opposed to a cassette tape) and *indexed* by number, making it ideal for remembering long, numbered lists and tables, or many smaller lists, or both: up to 10,000 basic items. Each basic item can, in effect, be elaborated with nooks and crannies [Hack #4], creating the potential for many more than 10,000 items.

In Action

You can think of the Hotel Dominic as a building with 100 floors, numbered from 00 to 99, each containing 100 rooms, also numbered from 00 to 99. In short, it's like a grid with 100 rows and 100 columns. The first room on Floor 95 would thus be numbered 9500. The next room along the hall would be 9501, then 9502, and so on. Figure 1-2 shows the first few rooms from the bottom floors of the Hotel Dominic, starting with the first floor, Floor 00. The hotel continues both up and to the right.

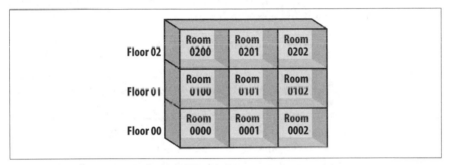

Figure 1-2. A few rooms in the Hotel Dominic

If you need to memorize a list with more than 100 numbered items, allocate an empty section of the matrix to that list. For example, to memorize the periodic table of elements, you could arbitrarily allocate rooms 8001 to 8116. Room 8001 would contain information about the first element, hydrogen, and 8116 would contain information about the element with the highest known atomic number, ununhexium (element 116).[1]

If you have memorized the 100 people/action pegs of the Dominic System, you *already* have everything you need to memorize the 10,000 rooms of the hotel. Room 8001 in the Hotel Dominic would be represented by whatever combination of person and action you have designated for HOOA. Four-digit numbers are easily memorized with the Dominic System, so all you need to do is associate hydrogen with the Dominic mnemonic for this number.[2]

In Real Life

Here's an example of how to memorize the first element in the periodic table, hydrogen.

As mentioned in the previous section, place hydrogen in Room 8001 of the Hotel Dominic (in other words, Floor 80, Room 01). The number 80, as mentioned in "Use the Dominic System" [Hack #6], is represented by Santa Claus (80 = HO, which reminds us of Santa's HO, HO, HO). In my personal list, 01 is that guy on the oatmeal box (01 = OA = oats), and his characteristic action is offering a bowl of oatmeal. Thus, 8001 is represented by Santa Claus offering me a bowl of oatmeal.

To link hydrogen to the picture, I imagine that the oatmeal is bubbling, and little bubbles of hydrogen gas are escaping from it and bursting into flame. It's then easy to remember other information about the element, such as its chemical symbol, *H*. I visualize a shiny *H* on Santa's forehead, as on the forehead of the "holographic" character Rimmer in the TV series *Red Dwarf*. I imagine that the shiny *H* reflects the flames from the exploding hydrogen bubbles, making it vivid and easier to remember.

Other features of the character, the character's action, and the room could be used to store other information about hydrogen, such as its atomic weight (1.008) and the year it was discovered (1766), both of which are four-digit numbers and also easily encoded with the Dominic System.

It's not as easy as playing foosball or watching a DVD, and experience shows that you'll need to refresh your memory periodically. However, "Dominate Your Memory" [Hack #8] provides a script to print up chunks of your personal memory hotel for easy review, and if you're studying for your chemistry exam or going on *Jeopardy!*, a mnemonic technique like the Hotel Dominic beats rote drill techniques and makes them cry.

You can use the Hotel Dominic to memorize lists *shorter* than 100 items, too. For instance, you can tuck the Universal Declaration of Human Rights (which has 30 articles) into rooms 8171–8200 and still leave plenty of room for new chemical elements to be discovered.

End Notes

1. Periodic Table Navigator. WebElements. *http://www.webelements.com/webelements/elements/text/periodic-table*.

2. This is similar to the way you would use Tony Buzan's SEM[3] (Self-Enhancing Master Memory Matrix) system to memorize the periodic table. In fact, the Hotel Dominic is an attempt to do with the Dominic System what Buzan has done with the Major System [Hack #5] in SEM[3]: *http://www.ludism.org/mentat/SemCubed*. However, the Hotel Dominic is nonproprietary and arguably easier to use.

See Also

- For more information on SEM3, consult Tony Buzan's books, especially *Master Your Memory*.

H·A·C·K #8 Dominate Your Memory

Use a Perl script to formulate items that match the 10,000 room numbers of the Hotel Dominic. Then, print the list as an aid for memorization and review.

"Visit the Hotel Dominic" [Hack #7] mentions a Perl script that will make memorizing large chunks of information with the Hotel Dominic method much easier and will also help you refresh your memory periodically. This hack contains that script.

With this new script, *dominate*, you will be able to print out as large a swath of the Hotel Dominic as you wish—hundreds or thousands of rooms—and mark it up with a pen or pencil, assigning each item you want to remember to a room. Then you will be able to review your marked-up version of the hotel at leisure and commit the items to memory.

The Code

Place the following Perl script in a text file called *dominate*:

```perl
#!/usr/bin/perl -w

$in_file = $ARGV[0];
$domstart = $ARGV[1];
$domend = $ARGV[2];

if ($domstart > $domend)
{
    die "Start number not less than or equal to end number\n";
}

open (IN_FILE, "< ./$in_file")
    or die "Couldn't open input file: $!\n";

$index = 0;
while (defined ($line = <IN_FILE>))
{
    $line =~ /([^;]*)\:([^;\n]*)/g;
    $domarray[$index][0] = $1;
    $domarray[$index][1] = $2;
    $index++;
}

close IN_FILE;
```

```
for ($domnum = $domstart; $domnum < = $domend; $domnum++)
{
    $domstring = sprintf "%0004.0d", $domnum;
   print "$domstring: ";
    $domstring =~ /(\d\d)(\d\d)/g;
    print "$domarray[$1][0]\,$domarray[$2][1]\n\n";
}
```

You will need to create your own datafile that contains your personal characters and actions that match the numbers in the Dominic System [Hack #6]. It must start with the character and action for 00 and continue through the character and action for 99. Each line must contain the character name, followed by a colon, then a space, and then the character's typical action. Colons must be used only to separate characters and actions; they cannot appear anywhere else in the file. If your text editor has a line number feature, you can use it to keep track of where you are in the file, such as line 1, which should contain the mnemonic for 00, or line 100, which should contain the mnemonic for 99.

Here are the last 29 lines of my *dominate* datafile, *dominic.dat*, corresponding to items 71–99:

```
$ tail -29 dominic.dat
```

```
Ray Charles: playing piano and singing ("GeorgiA")
George W. Bush: clearing brush in Crawford, TX
Graham Chapman: throwing open windows and exposing himself (scene from _The
Life of Brian_)
Gandhi: spinning on a wheel
General Electric (a robot in uniform): saluting
Smeagol-Gollum (_The Lord of the Rings_): falling into lava with the One
Ring
Gandalf the Grey (_The Lord of the Rings_): barring the way with staff
George Harrison: tapping foot and playing guitar
Ignatius Loyola: flogging himself (iGNatius)
Santa Claus: laughing until belly shakes (HO HO HO!)
Julia Sweeney: dressing gender-ambiguously (_God Said HA!_)
Humphrey Bogart: pulling the brim of his hat low
Hagbard Celine (_Illuminatus!_): piloting golden submarine
Howard Dean: whooping
Isaac Newton (long white wig): releasing helium balloons (Newtonmas, HE HE
HE!)
Homer Simpson (_The Simpsons_): gobbling donuts
Mercury (Greek god): flying with winged sandals (Hg = chemical symbol for
mercury)
Hermann Hesse (author of _The Glass Bead Game_): telling beads
Han Solo (_Star Wars_): firing blaster
Dr. No: manipulating controls in his secret headquarters
Neil Armstrong: stepping onto Moon
Norman Bates: stabbing someone repeatedly
```

David Sedaris (author): chasing rolling coins down the street (North
Carolina)
Nick Danger (Firesign Theatre, Phil Proctor): walking into a red sandstone
building (oof!)
Neelix (_Star Trek_): cooking alien food
Hiro Protagonist (_Snow Crash_): cutting someone to bits with a katana (Neal
Stephenson, author)
Morpheus (_The Sandman_): strewing sand (Neil Gaiman, author)
Nathaniel Hawthorne: tearing open his shirt to reveal a scarlet letter
Nick Nolte: falling into pool (scene from _Down and Out in Beverly Hills_)

Running the Hack

In "Visit the Hotel Dominic" [Hack #7], I suggested memorizing the Universal
Declaration of Human Rights[1] by placing it after the periodic table of elements in the Hotel Dominic. That hack placed the periodic table of elements in the hotel running from 8001 to 8170, with the 30 articles of the
declaration running from rooms 8171 to 8200.

However, since Santa (80 = HO!) is easier for most people to visualize than
Julia Sweeney (81 = HA!), for purposes of illustration only, we'll place the 30
articles in rooms 8071 to 8100. (Of course, you can place them anywhere
you wish.)

See the "How to Run the Programming Hacks" section of
the Preface if you need general instructions on running Perl
scripts.

If you have Perl installed on your system, to print out the "inhabitants" of
these 30 rooms, save the *dominate* script and the *dominic.dat* file in the same
directory, and then run *dominate* by typing the following command within
that directory. The first argument should be the name of the datafile (in this
case, dominic.dat), the second argument should be the starting room (8071),
and the third argument should be the ending room (8100):

```
perl dominate dominic.dat 8071 8100
```

If you're on a Linux or Unix system, you might also be able to use the following shortcut:

```
./dominate dominic.dat 8071 8100
```

The following is a set of results from an actual *dominate* run:

```
$ ./dominate dominic.dat 8071 8100

8071: Santa Claus, playing piano and singing ("GeorgiA")
8072: Santa Claus, clearing brush in Crawford, TX
8073: Santa Claus, throwing open windows and exposing himself (scene from
_The Life of Brian_)
```

```
8074: Santa Claus, spinning on a wheel
8075: Santa Claus, saluting
8076: Santa Claus, falling into lava with the One Ring
8077: Santa Claus, barring the way with staff
8078: Santa Claus, tapping foot and playing guitar
8079: Santa Claus, flogging himself (iGNatius)
8080: Santa Claus, laughing until belly shakes (HO HO HO!)
8081: Santa Claus, dressing gender-ambiguously (_God Said HA!_)
8082: Santa Claus, pulling the brim of his hat low
8083: Santa Claus, piloting golden submarine
8084: Santa Claus, whooping
8085: Santa Claus, releasing helium balloons (Newtonmas, HE HE HE!)
8086: Santa Claus, gobbling donuts
8087: Santa Claus, flying with winged sandals (Hg = chemical symbol for
mercury)
8088: Santa Claus, telling beads
8089: Santa Claus, firing blaster
8090: Santa Claus, manipulating controls in his secret headquarters
8091: Santa Claus, stepping onto Moon
8092: Santa Claus, stabbing someone repeatedly
8093: Santa Claus, chasing rolling coins down the street (North Carolina)
8094: Santa Claus, walking into a red sandstone building (oof!)
8095: Santa Claus, cooking alien food
8096: Santa Claus, cutting someone to bits with a katana (Neal Stephenson,
author)
8097: Santa Claus, strewing sand (Neil Gaiman, author)
8098: Santa Claus, tearing open his shirt to reveal a scarlet letter
8099: Santa Claus, falling into pool (scene from _Down and Out in Beverly
Hills_)
8100: Julia Sweeney, knocking something over ("Oh Oh!")
```

Your *dominate* script will print a double-spaced list so that you have room to annotate it by hand.

In Real Life

To remember the first five articles of the Universal Declaration of Human Rights, you might annotate them as follows, with a detailed visualization and a paraphrase of the relevant article:

8071: Santa Claus, playing piano and singing ("GeorgiA")
Two identical Santas sit side by side on a piano bench, each playing half the piano and singing in chorus. (Article 1: All human beings are born free and equal, and they should act in a spirit of brotherhood.)

8072: Santa Claus, clearing brush in Crawford, Texas
Santa is clearing brush in Crawford, knocking down old fences dividing people. (Article 2: Everyone is entitled to all the rights and freedoms in the Declaration, without any kind of distinction.)

8073: Santa Claus, throwing open windows and exposing himself (scene from The Life of Brian*)*

Santa throws open his windows and exposes himself to a death squad. Rat-a-tat-tat! He slams the windows shut. Whew! The windows are armored. (Article 3: Everyone has the right to life, liberty, and security of person.)

8074: Santa Claus, spinning on a wheel

Santa is chained to a spinning wheel and forced to spin, but he breaks his chains triumphantly. (Article 4: No one shall be held in slavery.)

8075: Santa Claus, saluting

Santa is being tortured on the rack; one of his elves frees him, and he salutes his liberator. (Article 5: No one shall be tortured or subjected to cruel, inhuman, or degrading punishment.)

End Notes

1. Universal Declaration of Human Rights. *http://www.un.org/Overview/rights.html.*

See Also

- The *dominate* script will also work for Tony Buzan's Major-System-based SEM[3] if you feed it the right data. For more information on SEM[3], consult Tony Buzan's books, especially *Master Your Memory* (David & Charles).

Memorize Numbers with Carroll's Couplets

HACK
#9

You can use a rhyming system of mnemonics by Lewis Carroll, author of the immortal "Alice" books and much nonsense poetry, to remember dates, phone numbers, and other numeric data.

In the 1870s, Lewis Carroll devised a mnemonic system for numbers that he called the *Memoria Technica*, after an earlier system. Carroll's system is little remembered by us postmoderns. Like today's more common Major System [Hack #5], it relies on converting numbers into consonants and filling them with vowels to make words; unlike the Major System, it uses rhyming couplets to help you remember the words that are created, instead of simply having you remember them "naked," and in this sense it is an advance on the former.

If you already know the Major System consonants, you could probably substitute them for Carroll's without too much trouble.

In Action

First, you need to memorize the number-to-consonant conversions shown in Table 1-5, which provides mnemonics for remembering the mnemonics.

Table 1-5. Number-to-consonant conversions

Number	First consonant	Second consonant	Mnemonic
1	B	C	First two consonants in alphabet
2	D	W	Duo; tWo
3	T	J	Tres (Spanish); see following note for an explanation for *J*
4	F	Q	Four; Quattuor (Latin)
5	L	V	*L* stands for *50*; *V* stands for *5* (Roman numerals)
6	S	X	SiX
7	P	M	sePteM (Latin)
8	H	K	Huit (French); oKto (Greek)
9	N	G	NiNe; *g* looks like *9*
0	Z	R	ZeRo

Carroll said his intent was to provide one common and one uncommon consonant for each number. He was a polyglot, so many of the metamnemonics involve number words in other languages; however, the only one that really doesn't make any sense is *J* for *3*. Carroll said it was the only consonant left after he filled in the rest of the table.

The next step is to convert the numbers you are trying to remember to a word or words and to make them the last part of a rhyming couplet. Carroll gives the following example to remember 1492, the year Columbus first came to America.

First, drop the *1* from *1492*; it's obvious Columbus didn't sail in 492 or 2492.

Next, convert 492 to a word using either of the pair of letters associated with each digit, like this:

Numbers	4	9	2
First option	F	N	D
Second option	Q	G	W

As it happens, the letters in the second row (FND) will form the word *found* nicely, but some other combination might have been used, such as QND for *queened* (as in "The pawn was *queened* when it reached the eighth row").

Carroll writes:

> The poetic faculty must now be brought into play, and the following couplet will soon be evolved:—
>
> "Columbus sailed the world around,
>
> Until America was F O U N D."[1]

Presto! Convert *FOUND* back to *FND* by extracting the vowels; then convert that to *492*, and you have it.

In Real Life

Carroll makes remembering numbers with rhyming couplets seem as easy as falling off a bicycle (they say you never forget how). But Carroll was an Oxford don, a mathematician, and a gifted poet. How easy is it for *us* to "bring our poetic faculties into play?"

I will now demonstrate the use of the method to remember the phone numbers of Powell's City of Books in Portland, Oregon (my favorite bookstore; visit it if you get a chance) and the Seattle branch of Ikea (where I do have to go periodically).

The phone number for Powell's City of Books is (800) 878-7323. We'll drop the toll-free 800 area code as being obvious, just as Carroll did with the *1* in *1492*. That leaves 878-7323. Using the mnemonics shown in Table 1-5, we can convert the phone number to letters, like this:

Numbers	8	7	8	7	3	2	3
First option	H	P	H	P	T	D	T
Second option	K	M	K	M	J	W	J

From these pairs of letters, we choose seven consonants that can form words:

H M K M T W T

And then the words themselves:

HAMMOCK MY TWIT

Note that I'm treating MM as a single 7 and CK as just K. I learned this consonant-melding trick from the Major System [Hack #5]; Carroll doesn't mention it. If you really do have a number with a double digit, such as 77 (MM), in it, either put a vowel between the consonants you use (as with MOM) or use two separate consonants (such as MP, as in *lump*).

Also, note that I had half an idea of what the final rhyme would look like when I selected the letters; as soon as I saw HMK, I thought of myself coming home from Powell's with an armload of books and lying down in a hammock to read them. The final result?

> Arms full of books, but you don't mind a bit?
> Lie down and read in the HAMMOCK, MY TWIT!

Now, for the Ikea Seattle store, whose phone number is (425) 656-2980. Again, I can omit the area code, because I know where the store is and what its area code is likely to be. That leaves me with 656-2980. Again, using the mnemonics shown in Table 1-5, we can convert to letters, as shown here:

Numbers	6	5	6	2	9	8	0
First option	S	L	S	D	N	H	Z
Second option	X	V	X	W	G	K	R

From these pairs of letters, I select:

S V S W N K R

SAVES A WANKER

I thought about rhyming *wanker* with *tanker*, *anchor*, and *Angkor*—all suggestive of the global reach of Inter IKEA Systems BV—but all the couplets I came up with were too long. Eventually, though, I devised this ditty:

> Look at all the cash and rancor
> That Ikea SAVES A WANKER!

Rude, eh? Don't worry; that makes it easier to remember.

Neither of these two rhymes took longer than a few minutes to create. However, they have stuck in my memory, suggesting that you use this method for data that's important to you, that you want to retain, and that you don't mind spending a little time learning. Carroll himself used his system to remember dates associated with various Oxford colleges, among other things; he would trot out the dates when showing guests around Oxford. Apparently, he used it to memorize logarithms as well.

A similar principle is at work in the mnemonic parody technique [Hack #10], which you can use to remember many more kinds of information than numbers.

End Notes

1. Collingwood, Stuart Dodgson. *The Life and Letters of Lewis Carroll*. Project Gutenberg. *http://www.gutenberg.org/dirs/1/1/4/8/11483/11483-h/11483-h.htm*. (This biography by his nephew includes *Memoria Technica* material in Chapter 7.)

See Also

- Facsimile of a handwritten *Memoria Technica* monograph by Carroll (*http://electricpen.org/CarrollMemoriaTechnica1.jpg*).
- Takahashi, Hisako. "Memoria Technica Japonica—A Study of Mnemonics" (*http://users.lk.net/~stepanov/mnemo/takahae.html*).

HACK #10 Tune In to Your Memory

Turn that song stuck in your head into a powerful tool to help you remember what you learn! This hack works especially well if you have a list of things to memorize.

It's common for people to hear a particularly catchy tune and hum it in their head for hours, or sometimes days. While this phenomenon can be annoying, it can also be used as a great tool for memorizing information. Making up a song or poem about a topic can be an extremely effective way to remember dates, lists of items, events and stories, and many other things.

This hack works in three ways to stick information in your mind. First, hanging information on a melody or rhyme scheme that you already know helps piggyback new information on information that you've already acquired. Second, remembering the rhythm of a tune or one or two rhyming lines can help bootstrap your memory; bringing one to mind will often bring up the rest of the associated information. Third, the active process of fitting the information into the tune causes you to concentrate on the information and turn it over in your mind, which also helps it to stick there.

There are a few different types of learning songs, and some may work better than others, depending on your own mental makeup or the information you're trying to memorize.

A *parody* is a song written using an existing song's tune, often satirizing or making fun of something. The parody might or might not play on the theme of the original song, but the new words often follow a rhyme or phonetic scheme similar to the original lyrics. They are also generally written based on a popular song rather than a folk or traditional tune. Matching the information you're trying to memorize to a song you already know, by theme or some other association, can further help you to remember the information.

If you need to learn a story, such as an event in history, putting the story to music with a *story song* will help you remember it. This hack has been used by people around the world for thousands of years, of course. Story songs are often similar to parodies, but may be more freewheeling and nonsatirical, and will probably use an original tune or traditional/folk tune.

List songs simply put a series of information to music or rhythm. They can be tricky to learn, depending on your list, but they can also be incredibly effective. List songs may take some time to memorize, but you won't soon forget them, and they are often faster to write than the other types. The keys to writing and learning a list song are rhyme and repetition.

In Action

To write a parody, begin with a topic you want to remember. Next, choose a popular song you know well and remember easily, or one that sounds like a word in your topic. This is best if you can transform the original lyrics into lyrics about your topic by changing only a few choice words. For example, if you're writing a song about baboons, pick an original song about a balloon, saloon, or something else that rhymes. How about "Up, up, and away, with my beautiful baboon"? Making humorous or absurd mental images makes things easier to remember, and when you're writing something funny, writing is a lot more fun. Often, once you start, the lyrics fall into place and you're laughing as you think of the next line. Continue working your information or story into the song until you get everything in, adding more verses if necessary.

Writing a story song to an original tune takes a certain kind of talent that the average person may or may not feel comfortable with. It's perfectly OK to use a traditional song here; it's also fine if your original melody isn't award-worthy, as long as you can remember it. If you choose a folk tune, you can either use the original lyrics as a base, as you would with a parody, or write your own from scratch. If you are using an original tune, you will have no base lyrics to work from, but coming up with your own lyrics can be half the fun. It can also make the memory stronger if you take time to craft all of the lyrics. Take the time to make it rhyme, too; rhyme and meter (*meter* is the rhythm of the words, as in poetry) are important to any song and will certainly help embed it into your mind. (Compare Lewis Carroll's couplets [Hack #9], which use a similar principle.)

List songs are often the easiest to compose. First, write out a list of information you'd like to memorize. This will give you a ready list when you need to pick the right word to rhyme or fit the meter. Next, find a tune that has either verses or repetitive sections you can repeat enough times to include your entire list. Start to sing the tune for your chosen song, and instead of the words, sing the list. You will probably have to add connecting words here and there and maybe at the end of lines to help the rhyme. If you can make the words on your list fit into the rhyme, however, it's more powerful. Making the list alphabetical can help you remember what comes next in the song. This works especially well if you have at least one word on your list for most letters of the alphabet (if not every one), such as the names of the states. You can also group information by geography (if you're memorizing countries in Africa, for example) or any other way that makes sense; this will aid your memory, too.

In all cases, physically writing or typing your work will help stick the information in your mind. Also, the more familiar your framework song or poem is, the easier it will be to remember its new words.

In Real Life

Parodies are particularly effective for remembering a group of related information or a story. The following parody, written by teenage girls, is not a mnemonic, but it's a good example of how to write a parody. "Negligee" is the story of a woman who buys an unfortunate piece of lingerie. It uses the tune to the Beatles' "Yesterday."

Negligee,	Yesterday,
I look stupid in this negligee,	All my troubles seemed so far away;
'Cause my butt sticks out this funny way.	Now it looks as though they're here to stay.
Oh, why'd I buy this negligee?	Oh, I believe in yesterday.
Negligee,	Yesterday,
I would like to take it back today,	Love was such an easy game to play,
But I threw the damn receipt away;	Now I need a place to hide away.
I'll have to keep this negligee.	Oh, I believe in yesterday.

If you were a child with access to a television in the '70s or '80s, you are probably familiar with *Schoolhouse Rock*. The producers of these animated shorts, which were shown between cartoons on network TV Saturday morning programming, knew well the teaching power of story songs. These songs, with accompanying animation, taught science, math, history, and English language skills to original catchy tunes, from "Conjunction Junction" to "My Hero Zero" to "The Preamble" (which set the preamble of the Declaration of Independence to music). Many other children's TV shows have also used this technique, notably *Sesame Street*; its efficacy is well documented.

The contemporary band They Might Be Giants have recorded many story songs that teach topics including mammals, James K. Polk (the 11th president of the United States), and the sun. Their most recent contribution is an album to help kids learn the alphabet, called *Here Come the ABCs*. There are songs about the letters themselves, such as "E Eats Everything," and songs about the alphabet in general, such as "Alphabet of Nations" and "Who Put the Alphabet in Alphabetical Order?"

The following example was recorded by They Might Be Giants and released as a single. Both of the core members of this band learned this song as kids from an album put out by Singing Science Records and remembered it all through their lives. When they discovered they both knew it, they wanted to record and perform it. It was an underground hit. These are the first few lines from "Why Does the Sun Shine?":

> The Sun is a mass of incandescent gas,
> A gigantic nuclear furnace!
> Where hydrogen is made into helium
> At a temperature of millions of degrees.

List songs have been recorded by many artists over the years, but all serve a similar purpose: to remember a long list that would otherwise be nearly impossible. List songs exist for memorizing the names of all the countries on Earth, the states of the U.S. and their capitals, and even the chemical elements. In "The Elements," comedian and MIT professor Tom Lehrer cheerily lists every name on the periodic table of elements to the tune of "I Am the Very Model of a Modern Major-General" from Gilbert and Sullivan's *The Pirates of Penzance*. Much like memorizing the digits of π, memorizing "The Elements" is a geek rite of passage.

Probably inspired by this, I wrote a song as a freshman in high school to teach my science class about the alkaline and alkali earth metals from the periodic table. Using little facts I found during my research, I wove them into the lyrics, sung to the tune of "Yankee Doodle." That was more than 15 years ago, and I can still remember it:

> Strontium turns the flame bright red
> So does rubidium
> Potassium turns the flame bright blue
> And cesium does too!
>
> Calcium is bright orange
> Barium is green
> Sodium is very bright and so it can be seen.
>
> Radium is radioactive—
> It gives you weird diseases.
> Calcium is found in milk
> And most of all hard cheeses...

See Also

- Unofficial Singing Science Records home page; *http://www.acme.com/ jef/science_songs*.
- Ask MetaFilter thread on catchy educational songs; *http://ask.metafilter. com/mefi/18731*.

—Meredith Hale

HACK #11   Consume Your Information in Chunks

Improve your short-term memory, your information processing, and your long-term memory by grouping the bits of data you come across into chunks.

Psychologist George A. Miller concluded in a classic 1956 experimental survey that human short-term memory can hold only seven items at a time, plus or minus two.[1] Short-term memory bears the same relation to long-term memory in humans that RAM does to mass storage in a computer: short-term memory, which is temporary, is the gateway to human long-term memory, which is semi-permanent. Short-term memory is also where information that is currently being processed is stored (such as a phone number you're calling). Thus, it's important for not only short-term memory itself, but also long-term memory and information processing, to maximize the ability to use short-term memory.

Recent research suggests the *magic number* that short-term memory can hold might be somewhat lower than seven, at least for intellectually demanding tasks. Researchers at the University of Queensland found that 30 academics given a task of analyzing statistical interactions among variables—a task at which they were already expert—did not perform better than chance at analyzing interactions of five variables in timed tests. Also, they were not only worse at analyzing four-variable interactions than interactions involving three or two variables, but less confident of their answers as well.[2]

Whether the magic number is five or seven, people normally find it hard to remember more than a few small *bits* of information. If they *recode* the bits by clustering them into larger, more meaningful *chunks*, however, they can remember many more of the bits. In the next section, we will show that you can remember a large number of *literal* bits (binary digits) by grouping them into more meaningful and comprehensible numeric chunks.

In Action

Here are 40 random binary digits. Examine them and spend as much time as you want memorizing them, then look away from this book and try to write them down. The only rule is that you may not convert them to another base, count them, restructure them, or use any other mnemonic trick to memorize them. You must memorize them by rote as you see them on the page. Are you ready? Go!

00111101111101111111000000100000010000011

How did you do? Probably not too well. If you did well, either you're a mutant with frontal lobes the size of a soccer ball, or you couldn't help but notice (for example) that the third group of 1s is seven bits long and is followed by six zeroes, then a 1 with five zeroes, and another 1 with five zeroes. In other words, you *chunked* them, you cheater!

Now, assuming you understand how to convert between binary (base two) and decimal [Hack #40], you can group these bits into bytes to produce five groups of eight bits, which you can *recode* to five decimal numbers, like this:

00111101	11110111	11110000	00100000	10000011
61	247	240	32	131

Look away from the book again. Try to write down the five decimal numbers and then convert them to their original binary form. Go!

How did you do at remembering the binary numbers? Probably better this time. While writing this hack, I noticed that I was spontaneously able to recall all five decimal numbers hours later without any formal mnemonic tricks. Chunking is simply a superior way for humans to process data! Miller writes:

> It is a little dramatic to watch a person get 40 binary digits in a row and then repeat them back without error. However, if you think of this merely as a mnemonic trick for exceeding the memory span, you will miss the important point that is implicit in nearly all such mnemonic devices. The point is that recoding is an extremely powerful weapon for increasing the amount of information that we can deal with. In one form or another, we use recoding constantly in our daily behavior; the kind of linguistic recoding that people do seems to me to be the very lifeblood of the thought processes.

Chunking is certainly central to many mnemonic hacks in this book, such as the Dominic System [Hack #6].

How It Works

The basis for the whole hack is *recoding* many small items that are difficult to distinguish (such as 40 bits) into a few distinct items (such as five decimal numbers). Five decimal numbers, even if they are not immediately meaningful to you, are few enough to retain in short-term memory. If some of them are meaningful to you, so much the better.

Rest assured that the technique applies to many phenomena other than binary numbers. Suppose you had a big pile of Scrabble tiles to memorize—say, 200 or so. If you could form them into words first, and then form the words into a sentence or paragraph, you'd have a much better chance of remembering the letters on the tiles.

In Real Life

When I was a broke psychology student, I often participated as a guinea pig in experiments at the Yale psychology and linguistics labs to earn pocket money. I had a few memorable experiences.

One experiment was designed to test the capacity of human short-term memory. A computer would flash strings of decimal digits rapidly on the screen, and the subject was supposed to type as many as she could back in. Although this was before I had made any serious study of mnemonic techniques, I instinctively chunked the digits into groups of two and three, effectively into numbers that I had made friends with [Hack #36]. I was able to beat the magic number of seven by about a factor of two without breaking a mental sweat. The experimenters seemed surprised and questioned me closely, apparently to determine whether duplicity was involved. I told them what I was doing, and they visibly relaxed; obviously, they knew about chunking, but apparently no one else who had participated in the experiment had used it.

End Notes

1. Miller, George A. 1956. "The Magical Number Seven, Plus or Minus Two: Some Limits on Our Capacity for Processing Information." *The Psychological Review*, 63. *http://www.well.com/user/smalin/miller.html*.

2. Halford, Graeme S., Rosemary Baker, Julie E. McCredden, and John D. Bain. "How Many Variables Can Humans Process?" (January 2005). *Psychological Science*. Abstract at *http://www.eurekalert.org/pub_releases/2005-03/aps-hmc030805.php*.

HACK #12

Overcome the Tip-of-the-Tongue Effect

Use what you can recall to help bootstrap your memory into remembering what you can't.

You are sitting with your friends, discussing the latest movie releases, when someone asks the name of a performer who starred in a recent film. Frustratingly, you can remember what she looks like, the fact that you saw her film from last year, and even that her name has three syllables and starts with an *A*, but the name just does not come to mind.

This experience, in which a memory seems to be "on the tip of the tongue," is exasperating if you're trying to remember a particular fact, but intriguing if you're interested in how memory works.

One of the most fascinating things about the tip-of-the-tongue state is that it demonstrates how sometimes we know that we know something, without actually being able to recall it. This is part of what psychologists call *meta-cognition*, which allows us to realize that we should keep trying even though our memories might be failing us at a particular moment. Much research has focused on metacognition and memory, because experiences like the tip-of-the-tongue state are relatively common in everyday life.

Studies have shown that tip-of-the-tongue states happen about once per week on average and get more common as we get older. Other research has focused on conditions that affect the likelihood of successful recall, suggesting some good techniques for overcoming tip-of-the-tongue when it occurs.

In Action

When people fall into a tip-of-the-tongue state, they commonly focus on the few relevant things that they can remember, hoping that the elusive fact will pop into their mind after the effort of increased concentration. A more successful technique is to try to recall as much information about the topic as possible, no matter how loosely it is related.

For example, in the situation described in the previous section, I might try to remember the plots and details of other movies I know the performer has been in, as well as what I was doing when I saw the original version of the film and who I was with. I could also try to remember what music was in the film, whether the actress has any brothers or sisters, and even which of my friends said she gave a good performance last time we talked about her.

If I could recall some aspect of the name (such as the number of syllables, or perhaps some of the sounds in the name), I could also try recalling words that sound similar, regardless of whether they are related in meaning to the thing I'm trying to remember.

As you work through these techniques, one of them will likely help you to recall the fact you are hoping to retrieve (in our example, the name of the woman in the movie). If you have other people to bounce ideas off, all the better, because it increases the chance that someone will be able to remember the answer.

You can use similar priming techniques to remember where you left misplaced objects.

How It Works

Memory is thought to rely heavily on a network of related mental concepts. The technique given here takes advantage of this network to make a difficult-to-recall fact more accessible to consciousness by activating as many related concepts in the network as possible.

One of the key concepts in psychology is *priming*, in which experiencing or thinking about one concept makes related concepts more readily available to the mind. For example, if the word *dinner* is shown to people in an experiment, they will react more quickly to words like *spoon* and *vegetable* than to words like *airplane* and *paper*, because words associated with food and dining are probably more closely and highly interconnected.

By thinking of as many related concepts as possible, you are increasing the activation in the area of memory that your target fact is connected to, thereby making it easier for your mind to lift the fact into consciousness.

The psychologist Endel Tulving proposed a related theory called the *encoding specificity principle*, which states that successful recall relies on the overlap between the thing you are trying to remember and the situation in which you first encountered it, and the cues or prompts that are available when you are trying to recall it.[1] The technique given here allows you to manipulate the context in your own mind to increase the chances of recall.

Just remembering related facts is only part of the process, however. Research has shown that hearing, reading, or thinking of similar-sounding words can also help overcome the tip-of-the-tongue state.[2] Models of language and memory suggest that meaning and word structure are stored separately, leading to the experience of remembering facts without being able to recall the word associated with them. In some cases, word structure is only partially remembered, so first letters or syllables are recalled, but nothing else. Priming seems to work as well for sounds as it does for facts and concepts, which is why you can remember the target word more easily by remembering words that sound like it.

End Notes

1. Tulving, E. 1983. *Elements of Episodic Memory*. Oxford University Press.

2. James, L.E., and D.M. Burke. 2000. "Phonological priming effects on word retrieval and tip-of-the-tongue experiences in younger and older adults." *Journal of Experimental Psychology: Learning, Memory and Language*, 26 (6), 1378–1391.

See Also

- Priming tutorial from Harvey G. Shulman; *http://www.psy.ohio-state.edu/psy312/priming.html*.

- Guide to tip-of-the-tongue experiences; *http://www.memory-key.com/EverydayMemory/TOT.htm*.

—Vaughan Bell

CHAPTER TWO

Information Processing
Hacks 13–18

Although memory is a core human faculty, and developing it will reward you well, as a literate human you still need to process recorded information, whether books full of text or digital files full of audiovisual data. How can you cope with the hurricane of information that pounds your eyes and ears every day?

This chapter will show you how to capture the best of the informational flood quickly, whether it comes from outside or inside your skull. It also will show you how to sort that information, structure it, and ultimately discard it from your life when you no longer need it.

Catch Your Ideas

#13 Good thoughts can come at any time. By recording them, you can bring them together and encourage your brain to give you more.

Interesting thoughts can come to you at any time. Perhaps you're getting groceries, in aisle A4, and suddenly you have an idea for a program you're writing. Or you're driving, and a point in an argument comes to you. Or you're in the shower, and you realize something about life.

But later, you simply forget. The very next day, you're tasked with writing that program, or giving your side in the argument, and you ask yourself, "Now what was it I was thinking?" Perhaps you are stuck living the same day over and over again. "Didn't I have a thought about a different way I could think and live?"

In this hack, you're going to collect your thoughts using a *catch*. This is not a simple diary; this is an advanced system for collecting every thought, from everywhere in your life, and bringing them together.

In Action

You will need some supplies:

- A ream of ruled paper
- A pen or pencil

Take a piece of paper, and prepare it like this. First, create three columns:

Subject

> The Subject column should be the leftmost column and be about an inch wide. This is the place where you will write the general subject of the idea you have. For instance, if you think a lot about C++ and you have an idea that's basically about C++ (rather than, say, math or philosophy), put "C++" in this column. You want to pick your subjects so that they are big and can hold a lot of related thoughts.

Hint

> The Hint column should also be about an inch wide and should sit next to the Subject column. This is where you will write a hint about how to place the idea within the subject. Perhaps it is a sub-subject—the name of a topic of interest within the subject—or a keyword that identifies a theme or context.

Idea

> The last column is for your idea. The idea that you will write down is the core idea for the thought you have.

Ideas are pregnant; they usually come with dozens of other details as well, which unfold from the main idea. You want to write out the main idea only. Write out enough of it so that you can trust yourself to unfold the rest of the ideas. Write one sentence, and write two only if you absolutely must. Skip the words *the*, *a*, and *an*. Those are signs that you are getting wordy, and that's not what you want to be. Just write the basic idea.

You're going to carry a piece of paper set up like this with you every day, for the duration of your catch. The next step is to collect related ideas. Periodically (perhaps once per day or once per week) peruse your pages of ideas. Ignore the actual ideas themselves; look instead at just the Subject column. Scan the list for recurring subjects.

Create new sheets of paper for the recurring subjects. For each subject, draw a sheet. At the top of the paper, title it with the subject. Then, form the three columns described previously. The first column (perhaps a centimeter wide) contains the idea number. Enumerate from number 1 (or 0, if you like). If this page is continuing a previously filled page, continue numbering from the previous sheet. The second column contains the hint. The third column contains the idea.

Transcribe the ideas that are on the same subject from the daily sheets that you carry around with you to the subject-specific sheets that you just made. This is the way that you can collect and refer to your ideas.

You can refer to them like this: *Subject Name, Idea #*. For example, you could call your 32nd idea on C++ *C++, 32*. When you can take the subject for granted, you can simply write `32 (pronounced *point 32*).

Computers and Tape Recorders

The obvious question is "What about computers? Can computers speed this up for me? Must we do this on paper?"

The answer is yes! Computers can probably speed this up for you.

With that said, your initial recording is probably going to be on paper. You don't want to lose *any* ideas! That means you have to be able to scribble something down *fast*. You can't be waiting for a computer to boot. You can't be moving a cursor around the screen. You can't be writing in some strange stylus pidgin. You just have to get the idea out there, so you can get back to whatever you were doing, as soon as possible. Transcribe to a computer if you have a good system.

"How about tape recorders?" I highly discourage using them. Transcription time goes through the roof because your interface is limited to fast-forward and rewind, and a muddy voice is harder to interpret than scrawl, in my experience. Also, you might forget to list a subject and a hint because you aren't prompted for them. Those two things are important, because they make it so that you don't have to recognize your thought during transcription and sorting, thereby saving you an amazing amount of time and mental energy. So, I discourage use of tape recorders.

All of these factors are changing. In the near future, there might be always-on computers with styluses, and voice recorders with excellent speech recognition. In 2005, though, I recommend capturing thoughts on paper.

In Real Life

When you have collected your ideas, you will perhaps think, "It's great that I have my thoughts here. Now, what can I do with them?"

Look for patterns. Think about how the ideas connect with each other. Look for holes in your thinking. Look for what you might be paying too much attention to. Reflect on what inspires you. Look for connections. You can do all of these things at once, by building a map of your related ideas. This is excellent fodder for a mind map [Hack #16].

Some things will become obvious to you immediately. You'll be surprised at the things you discover, once you put your thoughts together.

I suspect that paying attention to your mind causes it to give you more ideas by the same mechanism that dream recall works. If you've seriously investigated dream recall, you were probably surprised to find that the more you write down your dreams, the more you remember them [Hack #29]. I suspect also that the mind knows what kinds of things you want from it, based on what you are paying attention to. If you pay attention to your creative or interesting thoughts, your mind seems to decide to give you more of them.

Beware! If you implement this system, you may find that your problem is not too few creative thoughts, but too many. It's great to be creative and think interesting thoughts, but it can also be seriously immobilizing and can get in the way of day-to-day life.

See Also

- Kimbro, Lion. 2003. *How to Make a Complete Map of Every Thought You Think*. Published online at *http://www.speakeasy.org/~lion/nb*.

- WikiWikiWeb. "Is Anything Better Than Paper?" *http://c2.com/cgi/ wiki?IsAnythingBetterThanPaper*.

—Lion Kimbro

HACK
#14   **Write Faster**

Write smarter, not harder! The ASCII-based shorthand hack called Speedwords will not only enable you to write faster on paper without learning a special shorthand alphabet, but will also enable you to type faster in many word processors and text editors.

Dutton Speedwords is an artificial language [Hack #51] developed by Reginald Dutton in the early 1920s and improved over the following few decades. Dutton intended Speedwords both as an international auxiliary language like Esperanto, which could be written or spoken by people who did not speak the same native language, and as a shorthand system.

The advantage that Speedwords has over most other shorthand methods is that you do not need to learn a special alphabet to use it (as you would, for example, with the Gregg or Pitman shorthand methods). This feature not only makes Speedwords easy to learn, but also means that it can be typed, entered into PDAs with handwriting recognition systems, and generally used anywhere the Roman alphabet can be used. It's also great for quickly catching information [Hack #13].

In Action

This section contains a short Speedwords vocabulary, which should be enough to get you started.[1] The original Dutton Speedwords textbooks are long out of print, but there's plenty of material on the Web[2] if you want to go further.

One-letter Speedwords. If you just want to play around with Speedwords and give it a test drive, try learning the 27 single-character Speedwords in Table 2-1 and use them for a few days. Learning this list will probably take only a few minutes, and you might be surprised how natural it is to work them into your ordinary note taking.

Table 2-1. One-letter Speedwords

Speedword	Meaning	Notes
&	And	
a	To, toward, at	
b	But	
c	This, these	French *ce*
d	Of, from	French *de*
e	Am, are, (to) be, is	Latin *est*
f	For	
g	Them, they	
h	Has, have	Also used as auxiliary verb; for example, *G h go* = they have gone
i	In, within	
j	I, me	French *je*, Scandinavian *jeg*
k	That	As in "The movie that I am watching," not "That chair over there"
l	The	French *le*
m	With	German *mit*, Scandinavian *med*
n	No, not	
o	On	
p	Can, be able to, have the power or potential to	
q	Do...?	See the "In Real Life" section later in this hack
r	Will	Auxiliary verb; for example, *G r go* = they will go
s	He, him	*sh* = she
t	It	

Table 2-1. One-letter Speedwords (continued)

Speedword	Meaning	Notes
u	A, an, one	French *un*
v	You	French *vous*
w	Us, we	
x	If	
y	Was, were	
z	As, than, compared to	

Longer Speedwords. Table 2-2 contains the most important Speedwords for my own use, beyond the original 27. If you learn just these 100 or so words, you will greatly increase your writing speed.

Table 2-2. Frequently used longer Speedwords

Speedword	Meaning	Notes
ar	Friend	
at	Expect	
au	Hear	
az	Always	Contraction of *al oz*
azo	Never	See "Affixes" section later in this hack
bit	Piece, bit	
ce	Receive	
cer	Certain, sure	
ci	Decide	
da	Give	Latin *dare*
dir	Direction	
dok	Document	
du	Continue	
dy	Since	French *depuis*
eb	Even	German *eben*
ef	Efficient, capable (effective)	
eg	Equal	
en	Attention	
ep	Place, position, location, to put	
er	Person	
es	Estimate, guess	
et	Small	Booklet, islet

Table 2-2. *Frequently used longer Speedwords (continued)*

Speedword	Meaning	Notes
fas	Easy	
fn	Find	
fy	Cause, reason	
ga	Complete(ly)	German *ganz*
gar	Keep	
ge	Together, join	
ha	Own, possess	
hab	Ordinary	
haz	Chance, luck	Hazard
ig	General (common)	
in	Between, among	Latin *inter*
it	Tool	
ite	Travel	
iv	Associated with	
je	Every	German *jeder*
jm	Everything	Contraction of *je om*
jr	Everyone	Contraction of *je er*
ke	Credit, due	
kon	Agree	
kre	Belief, believe	
ku	Enclose, include	
la	Big	
las	Permit, let	German *lassen*
lib	Free, release	Free as in freedom
lim	Boundary, limit	
lu	Month	
lut	Contest	
ly	Long and thin	
mem	Memory, remember	
mir	Wonder(ful)	
miu	Minute	60 seconds
mot	Word	
nar	Story	
ne	Take	German *nehmen*
nes	Need, necessary	
nm	Nothing	Contraction of *n om*
no	Look at	

Table 2-2. Frequently used longer Speedwords (continued)

Speedword	Meaning	Notes
nr	Nobody	Contraction of *n er*
nu	Now	Dutch and Scandinavian *nu*
ny	Almost, approximately	
ob	Get	
op	Against, opposite	
ord	Order	Opposite of disorder
ov	Over	
oz	Happen	
pe	A while, period	
pin	Point	
pl	Pleasure, to please	
por	Important	
pru	Prove	
rap	Fast	Rapid
ro	List	
ry	Building	
ser	Look for, search	
sev	Divide, division	Sever
sig	Meaning, to mean	
so	So, such	
sol	Only, alone	
stu	Learn, study	
su	To improve, better	
suk	Succeed, success	
sy	Science	
tru	Through	
ub	Favorable	
us	Use	
uz	Once	
va	War	
vo	Willing	
vot	Choose	
vu	See, sight	
we	Purpose, intend	
zi	Because	

Some auxiliary verbs. Table 2-3 presents a few useful verbs that it is easy to conflate, grouped for easy reference.

Table 2-3. Auxiliary verbs

Speedword	Meaning	Notes
yr	Would	Will (past tense)
yp	Could	Can (past tense)
debi	Should	
ypi	Might	

Affixes. Dutton Speedwords uses *affixes* (prefixes and suffixes) to extend its basic vocabulary. Table 2-4 lists only the most important affixes.

Table 2-4. Affixes

Affix	Meaning	Examples
-a	Unfavorable	en = attention, ena = worry; pro = promise, proa = threaten
-b	Possibility	krc = believe, kreb = credible
-c	Collective	ci = decide, cic = committee; on = man, onc = community
-d	Passive	ri = write, rid = written
-e	Intensive	ny = near, nye = next; Ja = soon, Jae = immediately
-f	Causative (after vowels)	ta = late, taf = delay
me-	Comparative	la = large, mela = larger
my-	Superlative	la = large, myla = largest
-n	Negation	ok = correct, okn = incorrect
-o	Opposite (after a consonant)	up = up, upo = down; ov = over, ovo = under
-p	Place	au = hear, aup = auditorium
-r	Person	ny = near, nyr = neighbor
-st	Professional	ju = to judge, just = a judge
-t	Diminutive	bo = tree, bot = a plant; nav = ship, navt = boat
-u	Favorable	haz = chance, hazu = lucky or fortunate
u-	Present participle	pu = think, upu = thinking
-x	Opposite (after a vowel)	bi = life, bix = death
y-	Past tense	pu = think, ypu = thought (but h = have, hy = had)
-y	Causative (after consonants)	bix = death, bixy = kill
-z	Plural	bu = book, buz = books

Many of Dutton's affixes are idiosyncratic; this is sometimes listed as a major drawback of the language. The idiosyncratic quality of Dutton's usage is a problem only if you are using Speedwords as an international auxiliary language, not if you are merely taking notes for yourself.

How It Works

The basis for Dutton Speedwords is a principle known as Zipf's Law. Informally put, Zipf's law states:[3]

> In most languages, there are a few very common words (such as *a*, *and*, *of*, and *the*) and a large number of uncommon words (such as *spontaneity* and *forensics*).

Zipf's Law applies not only to human languages, but also to computer languages, colors in computer graphics files, audio frequencies in sound files, and so on. This fact is the basis for most compression techniques, which reserve short encodings of one or two characters for common items and use long encodings for uncommon items.

Although the English language has already evolved a kind of compression via Zipf's Law (*of* is already shorter than *spontaneity*), Reginald Dutton engineered a language that would compress language even further.

In Real Life

A language is more than its vocabulary. You can't use a language without knowing its grammar as well. Fortunately, the grammar of Speedwords is very simple.

You can use most Speedwords as nouns, adjectives, or adverbs without changing them. This is similar to languages such as Chinese, which do not distinguish between parts of speech in this way.

 Actually, Dutton was inconsistent on this point; sometimes he did distinguish between these parts of speech, but that doesn't mean you need to.

Speedwords uses compound words, such as *ca* + *dor* (i.e., *room* + *sleep*), which means bedroom. Note that the way in which Speedwords forms compounds such as *cador*, literally *roomsleep*, is the reverse of the way in which English speakers would expect to form them (*dorca* or *sleeproom*).

The prefix *y-* indicates past tense, and the letter *r* indicates future tense: *j sa* means "I know," *j ysa* means "I knew," and *j r sa* means "I will know." When standing alone, *y* is the past tense of *e*: *sh y fe* means "She was happy."

The continuous or progressive verb is also expressed with *e*; for example, *j e sa* means "I am knowing."

The English verb *do* is not translated into Speedwords when it is redundant: *j n sa* means "I (do) not know." The word *to* in the infinitive of verbs is also considered redundant; thus, *sa* means "know" or "to know."

Pronouns are not inflected to indicate case: *w* means "we" or "us." Possessive pronouns are created with *-i*: *ji* means "my," *si* means "his," and *shi* means "her(s)."

 Possession is indicated for nouns with an apostrophe: *ji ar' buz* means "my friend's books."

Nouns don't specify plurality when it is already indicated in some other way (e.g., *ji 5 bu* means "my five books"), but when plurality isn't already specified, use *-z* (e.g., *ji bu* means "my book," and *ji buz* means "my books").

Sentences that describe the existence of an item do not contain the word *there*: *E 3 bu ir f v* means "(There) are three books here for you."

Yes-or-no questions are simply prefixed with the letter *q*. For example, *v pu k…* means "you think that…" and *q v pu k…* means "do you think that…?"

For most other questions of grammar, you can treat Speedwords as a "relexification" of, or code for, ordinary English. You might not always write grammatical Speedwords, but if you're only writing for yourself, this should not be a problem.

Q v nu h ci ri i rapmotz? J n es k t r ne v u lo pe.

End Notes

1. Harrison, Rick. 1994. "Language profile: Speedwords." *http://www2. cmp.uea.ac.uk/~jrk/conlang.dir/Speedwords.overview* (apart from my personal use of Speedwords over the past 12 years or so, most of this hack was drawn from Rick Harrison's 1994 overview of Speedwords, which he generously placed into the public domain).

2. Mentat Wiki. "Shorthand System." *http://www.ludism.org/mentat/ ShorthandSystem* (contains many Speedwords links, as well as other ASCII shorthand systems).

3. National Institute of Standards and Technology. "Zipf's Law." *http:// www.nist.gov/dads/HTML/zipfslaw.html*.

HACK #15 Speak Your Brain's Language

To absorb new information and to assimilate it quickly and effectively, use learning-style theories to understand your brain and what makes it function best.

If you have to drive somewhere new, how do you figure out how to get there? Perhaps you like to consult a web site and get a step-by-step list of driving directions that details each turn and street name. Maybe your dad prefers looking at a map and tracing out his route there. If you asked your friend, she'd tell you she likes to call someone and ask for directions, including landmarks and possible pitfalls she might encounter on the way. Maybe you've even had arguments about this, with each side claiming the only "good" way to be sure you get there, and secretly thinking that the other ways are for idiots.

This argument happens because each person has a different *learning style*. A learning style is a way of taking in and assimilating information, and different people's brains do this in different ways. So, each method of finding out how to get where you're going might be right for the person who favors it, and for him it might really be idiotic for that person to use a method that's less effective.

If you can tune in to the best way for your brain to learn, you can apply that knowledge intelligently to learn faster and retain more of what you learn. Knowing a little about learning styles will help you "speak your brain's language" so that it works better for you.

Learning Style Theories

There are many, many theories about how people learn. This hack discusses two that have many supporters and that I've found to be useful. Furthermore, they mesh well so that you can combine them synergistically; one is about how the information's format affects the brain's ability to take it in, and the other has to do with how the brain assimilates new information.

The VARK system. Neil D. Fleming and Colleen Mills developed the VARK system to describe different ways people absorb information.[1] VARK stands for the four types of learning defined by the theory: visual, aural, reading/writing, and kinesthetic. Most people can take in information through all of these *channels*, but each person usually has one preferred format:

Visual

Visual learners understand when they can see a representation of the material. They learn best from maps, charts, pictures, diagrams, video, and other nonverbal formats. They don't retain information well from speeches, books, or anything else that's text-heavy. For example, they might remember phone numbers by picturing the pattern their fingers make on the keypad as they dial.

Aural

Aural learners like to hear information and can take it in easily when they can remember what it sounds like. They learn best from lectures and discussion. For example, they often remember the words to songs and might remember phone numbers by the tones they make when they're dialed.

Reading/writing

Reading/writing learners love words and text. They learn best by reading about something and then writing a summary. For example, they might remember phone numbers by writing them down or making a code out of the digits.

Kinesthetic

Kinesthetic learners are tuned into their bodies. They like information they can use and remember information by remembering where they were at the time or how it felt to perform a new skill. They learn best by practicing and might remember phone numbers by making them into a little finger dance.

Honey and Mumford's learning styles. You can complement the VARK system with the learning styles defined by Peter Honey and Alan Mumford.[2] This schema tells less about the format of incoming information and more about how people assimilate the information once they receive it.

Honey and Mumford define four styles or roles that indicate a preferred approach to new information:

Activist

Activists like to take an active role in learning—no surprise! They learn best when they're involved in a discussion, leading a group, performing a task or game, or thrown into a new situation that they have to figure out. They like new ideas and new projects but might have trouble with implementation and detailed planning.

Activists don't learn well from long lectures or written instructions, especially if they have to work on them alone.

Reflector

Reflectors like to examine new information carefully, from all sides. They observe rather than take action at first, and they seek many different angles and viewpoints before expressing an opinion. They like to review and make sure they have all the data they need before proceeding.

Reflectors don't work well under deadlines, and in a group setting they'd much rather produce necessary analysis than lead and delegate.

Theorist

Theorists work to fit new information into a rational scheme. They're most comfortable with ideas they can adapt and integrate into logically sound theories, and they like to learn in an orderly, step-by-step fashion.

Theorists interact with information, prefer orderly tasks with a clear goal, and often probe for the ideas under other ideas to test their theories. They don't do well when asked to handle subjective information rather than hard facts, or when they have to perform "arbitrary" tasks whose function they don't understand.

Pragmatist

Pragmatists like to know how they can use new information. They're practical and always want to see how something will fit into their established methods or how new ideas will benefit them in some concrete way. They usually want to "cut to the chase" or ask about the bottom line, and they're impatient with abstract or theoretical discussion.

Leveraging Your Learning Styles

To repeat, most people use all of the VARK *channels* sometimes, and most people have situations where they use each of the learning styles, but almost everyone has one or two that fit most closely. After you identify your best learning format and your preferred learning style, you can begin tailoring your learning situations to fit.

There will always be situations in which you can't change some features of the information; if you're at work and need to learn a new process from a verbal presentation, you probably can't demand that the presenter turn it into a simulation game or documentary film. However, you'd be surprised at how often you can "translate" for yourself with a little ingenuity.

VARK types. Here are some suggestions about altering an information format to resemble more closely the type of information your brain prefers. Just find your type and try some of the ideas listed under it.

Visual

Draw pictures and diagrams, visualize processes and objects, look at photos and videos when possible, use colored markers and other tools to make text visually interesting, and replace words with symbols in notes. Don't forget to practice translating the other way to put your pictures back into words, since you'll usually be required to articulate your ideas that way verbally. Visualization [Hack #34] and diagramming [Hack #16] should work well for you.

Aural

Repeat facts to yourself out loud, attend lectures and presentations, discuss things with other people, use a tape recorder to catch notes from yourself or another speaker, take written notes sparsely using outlines or key phrases (possibly leaving space to fill in later as you remember information afterward), and concentrate on listening to the speaker instead. Try setting information to music [Hack #10].

Reading/writing

Seek out reference books and other textual material on your subject, write lists and outlines to remember things to do, take detailed notes, rewrite information in your own words, write text descriptions of pictures and diagrams, and explain your ideas to someone else in email or on a blog. Idea catches [Hack #13] and dream journals [Hack #29] will be useful for you.

Kinesthetic

Create sense memories about information (imagine how something might smell or feel), participate in demonstrations or tests, imagine what your body does in performing a new task, think up examples for ideas, role-play, and link ideas to objects and motions. You'll get good use out of sense-based training [Hack #74] and memory journeys [Hack #3].

Honey and Mumford learning styles. After you've made the information more palatable to your brain, use your best learning style to help retain it. Select your style from the following list and try some of the activities associated with it:

Activist

Jump into new experiences feet first, form a task force, volunteer to lead a group or discussion, and trade detail-oriented tasks for "big-picture" tasks when possible. Try *making a game* out of what you have to learn or do.

Reflector

Talk to other people, review what you know, assimilate information into graphs or tables, make sure to give yourself plenty of time, and see if you can get someone else to run through a test situation while you observe. When something works, go back and see how you did it [Hack #32].

Theorist

Make sure you understand the goal or purpose of what you're doing, get the larger context of what you're learning and not just bare processes, and draw diagrams of how ideas are related to each other. Never be afraid to ask questions [Hack #55] until you understand the topic thoroughly.

Pragmatist

Figure out how new ideas will profit you, find a model you can copy, and seek feedback to make sure you're not wasting your time and effort. Learn some general time-saving techniques, such as how to test for divisibility [Hack #37].

In Real Life

My first experience with applying learning theory came when I started studying a lot with other people in college. I was amazed at how many detailed notes other people took; my note taking was always very sparse. I was determined to be more diligent about taking copious notes, but I found that the more notes I took, the less information I retained.

I had been instinctively playing to my strengths; I'm a very aural learner, so I retained a lot more information if I listened carefully to a lecture instead of putting my attention on note taking. Further evidence about my proclivity comes in the fact that I've been notorious all my life for remembering song lyrics accurately, sometimes after having heard the song only once.

Ron, the primary author of this book, is so clearly a theorist that it made me laugh out loud when I was researching and reading descriptions of this type. You might also notice lots of hacks here that suit reader/writer learning; Ron is never without his notebook, and he has kept a journal assiduously since he was 15 years old. He also said that one of the things he liked most about creating this book was solidifying his understanding of the hacks by writing them.

It has been very interesting for Ron and me to discuss this hack, since the style differences it details explain several things we've frequently misunderstood about each other for years. This is another, secondary use for this hack: it will help you understand your family, friends, and co-workers better, and when you need to communicate with them, you'll have a better idea of how to do it effectively.

End Notes

1. Fleming, N.D., and C. Mills. 1992. *Not Another Inventory, Rather a Catalyst for Reflection*. To Improve the Academy, 11, 137–155.

2. Honey, P., and A. Mumford. 1992. *The Manual of Learning Styles*. 3rd Ed. Peter Honey Publications.

See Also

- More information about Fleming's VARK system is available at the VARK web site (*http://www.vark-learn.com/english/index.asp*), including testing and teaching materials, further articles, and much more.

- If you'd like to learn more about learning styles generally, Greg Kearsley's Theory Into Practice Database (*http://tip.psychology.org/index.html*) is a compendium of 50 theories about learning and instruction, which you can browse individually or by related groups.

- Many, many web sites will allow you to answer some questions that help determine your learning style. Not all of them are cost-free to use, but many are. A few minutes in the search engine of your choice should turn up some options.

—*Marty Hale-Evans*

HACK #16   Map Your Mind

Collecting and connecting related ideas reveal patterns that stimulate new thought, as well as contradictions to be resolved.

Are your thoughts organized? Most of the time, people live in a river of thoughts and sensations, like the story line of a movie. Our thoughts arrive as events in a sequence.

But we can map out our thoughts on a plane. When we see them side by side, we can compare them with one another and organize them. Observing the whole picture, all at once, we make startling realizations and discover an order to our thoughts, a top-down understanding of them.[1]

Alternatively, we find a disorder, and gain insight into tensions and confusions in our life. As soon as we see them clearly, though, our mind starts cranking away, working to resolve them—or, at the very least, to understand the subtlety behind the tension.

In Action

Mind mapping begins with collecting thoughts into a *source list*: a list of ideas you start out with and that you're going to map. It's useful to separate assembling your mind map from collecting your source ideas.

Creating the source list. Your source list can come from free writing, from a catch [Hack #13], or even from a chat transcript. What's necessary is to turn the source into a list.

Let's start with free writing. Think of a subject you think about a lot. Situate yourself in front of a keyboard, close your eyes, and then type out everything that comes to you about the subject. You can try to focus on one topic, or fan out to just about everything important to you. Whatever you think about will appear in the mind map.

When you are done, read what you wrote. Wherever you spot a complete and distinct idea, enumerate it. Enumerate ideas, not sentences. Three sentences on the same idea receive just one number. One sentence with three ideas in it receives three numbers.

Suppose you had written the following in stream-of-consciousness free writing:

> I think a lot about programming. I keep wondering, what about block-level design patterns? I've noticed that people who are starting to program don't know how to hook from "I'm visiting every member of a two-dimensional array" to "I need two nested for loops." What can we do about that? I think we can make "Block Level Design Patterns." They're too elementary to be noted by the sophisticated Design Patterns community, but I think they would be useful nonetheless. We could show patterns of exception use, patterns of conditional loops, and things like how to articulate decisions into variables, and we could explain all the trade-offs involved in these things.

Enumerated, it would look something like this:

> I think a lot about programming. I keep wondering, what about (#1) block-level design patterns? I've noticed that (#2) people who are starting to program don't know how (#3) to hook from "I'm visiting every member of a two-dimensional array" to "I need two nested for loops." What can we do about that? I think we can make "Block Level Design Patterns." They're (#4) too elementary to be noted by the sophisticated Design Patterns community, but I think they would be (#5) useful nonetheless. We could show (#6) patterns of exception use, (#7) patterns of conditional loops, and things like (#8) how to articulate decisions into variables, and we could (#9) explain all the trade-offs involved in these things.

Now, draw up the source list:

1. Block-level design patterns
2. Block patterns: for people starting to program
3. Block patterns: visiting over 2-D array needs nested loops
4. Block patterns: too elementary for the Design Patterns community
5. People: useful for learning
6. Block patterns: exception handling
7. Block patterns: conditional loops
8. Block patterns: decision into variable
9. Block patterns: explain trade-offs

Note that we list the number, a context hint, and a few words to describe the idea. The context hints are just like the context hints in a catch.

Now that you have your source list, it's time to assemble your mind map.

Assembling the mind map. Take a few ideas from your source list, and write them on a piece of paper; take a few more, and map them out too; and so on, until you've mapped them all. Continuously, in the background, think about the map as a whole. By the time you reach 30 to 50 items, the map should be fairly interesting, with several clusters.

Put your first idea anywhere on the page. When you place your second and third ideas, think about the relationship between the ideas. Place related ideas *nearby*, and put unrelated ideas *far away*. When you place an idea, just write the number and one to three words.

Make important ideas large. Make unimportant ideas small. You can even abbreviate small ideas to just their number.

After a while, you'll have interesting clumps of ideas. When you can name the theme relating ideas together, write down the name in big letters between the ideas. I call these names *magnet words*, because they represent what pulls ideas together.

With time, you'll need more than just distance to show structure. Introduce *lines*. Connect related major ideas by line to see structure. Vary the length, darkness, path, and arrangement of the lines.

I like to use color. I recommend four-color pens, if you are doing this with paper. (Bic makes some that are commonly available.) You can establish any sort of coloring convention; I tend to use blue for structural elements (lines, magnet words, etc.) and black for ideas.

As you draw, you will naturally pay more and more attention to structure. You might need to redraw portions of the map a few different ways to see what works right and feels right.

Ideally, with few exceptions, your structure will give each idea one obvious place for it to go. If you find that a pair of ideas meets and that they are really the same idea, just expressed in different ways or from different contexts, this is a good sign. It means your thinking is synchronizing different contexts that share something.

> If you find contradictions exposing themselves, this, too, is good.

How It Works

Mind mapping works because it challenges you to place every thought with respect to all your other thoughts. Whenever you place a thought onto the map, you ask yourself, "Now, how does this idea fit in with all the other ideas?"

But it's faster than comparing one thought to every other thought on the list. When you look at your map, you ask yourself, "Where should this go?" Since the map is spatially arranged, you move the thought to whatever area feels "warmer." When you find the general place for your thought, you need to compare it only against its spatial and structural neighbors. You can ignore the other thoughts, for the most part, because they are far away. In this way, you efficiently place thoughts.

When you've placed all your thoughts this way, in relationship to all the other thoughts and with the structure to see the ideas, you end up with a very nice top-down organization of your thinking.

In Real Life

Figure 2-1 shows the mind map created from the source list in the previous section. Just like the source list, each idea in the mind map consists of its number, a context hint (which can come from spatial position or links to other ideas), and a few words to describe it. The intention is to be able to reconstruct the stream of thoughts you started with (in this case, the paragraphs of the example beginning with "I think a lot about programming") and to understand how each of these thoughts relates to the others.

A mind map created by another person can only be a pale example, though; if you create your own, you'll really see how useful mapping your mind can be.

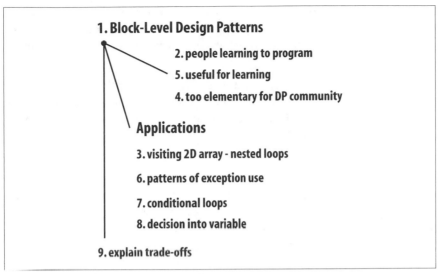

1. **Block-Level Design Patterns**

    2. people learning to program

    5. useful for learning

    4. too elementary for DP community

Applications

    3. visiting 2D array - nested loops

    6. patterns of exception use

    7. conditional loops

    8. decision into variable

9. explain trade-offs

Figure 2-1. A mind map for block-level design patterns

End Notes

1. Buzan, Tony, and Barry Buzan. 1993. *The Mind Map Book*. Penguin Books USA.

See Also

- Kimbro, Lion. 2003. *How to Make a Complete Map of Every Thought You Think*. Published online at *http://www.speakeasy.org/~lion/nb*.

—Lion Kimbro

 ### HACK #17 Build an Exoself

Take a Hipster PDA, combine it with a pocket countdown timer called the MotivAider, and gain better control of your thoughts, emotions, and activities.

Science fiction writer Greg Egan explores the concept of the *exoself* in some of his novels. In *Permutation City*, he defines it as "sophisticated, but non-conscious, supervisory software which could reach into...brain and body and fine-tune any part of it as required."[1] In a later novel, *Diaspora*, he describes the exoself's *outlook* component: "software that could run inside your exoself and reinforce the qualities that you valued most, if and when you felt the need for such an anchor."[2]

Pretty exciting! I'd give a lot for a mental exoskeleton that I could program one day to make me less lazy the next day, and keep me from getting sucked into a cult or pyramid scheme the day after that. Needless to say, however, we lack the technology necessary to build a true exoself today. Of course, it's a lot easier to reprogram yourself when you've been uploaded into a computer, as in Egan's fiction.

This hack creates a simpler system for repatterning your thinking, by using a Hipster PDA (made from a deck of index cards) and a periodic alarm device such as the MotivAider. Compared to Egan's fictional exoself, it's almost embarrassingly primitive—but it's a start.

In Action

Here's how to design an exoself with today's materials. Note that the design is flexible. Many components have substitutes. If we ever develop real exoselves, they'll be more complex than today's most powerful computer, so feel free to elaborate on this extremely basic design.

Stack. Bind together a stack of index cards with a binder clip or rubber band. Merlin Mann of the 43 Folders productivity blog made this design popular as the "Hipster PDA,"[3] but the stack is more structured than the loose bundle of notes that that term implies. Here are a couple things no good stack should be without:

Instruction cards
> As you encounter high-priority to-do-list items, place them at the top of your stack, with the most important item on top, followed by other items in decreasing priority. For example, the top instruction card might read, "Finish writing paper." The one beneath it might read, "Celebrate writing paper." After you "execute" the top instruction card, discard it and start the one beneath it. Thus, in this case, first you'd write your paper, then go out for a little R&R. Your stack works much like the instruction stack in a computer; only with this hack, you're programming yourself.

To-do list
> You should print this four-up and duplex for compactness, if possible, and then bind it to your stack with rubber bands, a binder clip, or something similar. The stack proper contains tasks you intend to do today, and the to-do list contains tasks you plan to do eventually. The goal is to gradually transfer all of the items on your to-do list into your stack and "execute" them. An intermediate staging area between the to-do list and instruction cards might be *calendar cards*: one for each day in the next week and each month in the next year. You can bind these anywhere in your stack that you think is appropriate.

To-do-list software. Use any to-do-list software with which you're comfortable. I use Emacs PlannerMode.[4] The important thing for the purpose of this hack is to be able to print out your to-do list so that you can bind it together with your stack.

Your outlook. Not to be confused with the bloated software product of a similar name, this is the part of your exoself that maintains your clarity and focus. In the back of your stack, insert cards with motivational quotations, short mental exercises to maintain mood and rationality (such as those in the emotional ABCs [Hack #57]), or reminders and heuristics (such as "Remember to buy my wife presents, just whenever"). Consult your outlook when you feel stressed, or simply whenever it occurs to you.

MotivAider. The MotivAider[5] is a small device, much like a pager, that you can set to vibrate at set intervals to keep you focused on whatever behavior you wish to associate with it. You can put this silent countdown timer into your pocket or clip it to your belt, and then set it to intervals ranging from one minute to one day. It will vibrate at the specified interval, then immediately reset itself, so if you set it for 45 minutes it will vibrate silently at 45-minute intervals. Later models can vary the interval randomly.

The MotivAider is a sublime tool. Some people use it to remind themselves to smile or not to slouch. You will use it to remind yourself to consult your stack. "Program" yourself to consult your stack whenever the MotivAider goes off, do what's on top, then position the next card. Write new cards as needed, offload long-term to-do items to your to-do list, and sometimes move them the other way.

Backup system. Scan the cards in your stack, plus any other paper notes you have, and then save them to hard disk with your to-do list and other critical personal data. Check your critical data into CVS or Subversion and find a friend with whom you can exchange remote backups. Encrypt sensitive data so that remote sites can't poke around in the files. If you enable secure remote access to your repository, you can access your critical personal data from anywhere in the world that has an Internet connection.[6]

In Real Life

I haven't yet scanned all my notebooks and index cards and checked them into Subversion; however, I have done everything else in this hack, and it keeps me on track better than anything else I've tried.

Here are some examples of the instruction cards I have written for my exoself:

- Card 1: Finish your task review. Card 2: Surf your del.icio.us inbox as a reward for finishing your task review. (This pair of cards is an example of snapping yourself to attention [Hack #74].)
- Do the next three cards now! (You can make using your exoself fun.)
- Do this card; then stick it back onto the stack, five cards down. (This card creates a periodic reminder.)

Sure, it's more like a child's crayon drawing of an exoself than the real thing, but it could be your head start on the Singularity. (See *http://www.ugcs.caltech.edu/~phoenix/vinge/vinge-sing.html* for more on the Singularity.)

End Notes

1. Egan, Greg. 1994. *Permutation City*. HarperPrism.
2. Egan, Greg. 1998. *Diaspora*. HarperPrism.
3. Mann, Merlin. 2004 "Introducing the Hipster PDA." *http://www.43folders.com/2004/09/introducing_the.html*.
4. EmacsWiki. 2005. "Planner Mode." *http://www.emacswiki.org/cgi-bin/wiki/PlannerMode*.
5. MotivAider; *http://habitchange.com*. (You can order a MotivAider here. The site also contains downloadable documentation for the device.)
6. Hess, Joey. 2005. "Keeping Your Life in Subversion." *http://www.onlamp.com/pub/a/onlamp/2005/01/06/svn_homedir.html*.

See Also

- Mann, Merlin, and Danny O'Brien. 2005. "Focus: Or, Why Your Web Browser Needs a Hypothalamus." *Make*, Volume 2 (O'Reilly). (Ruminations on how to remind yourself to stay on track.)
- MarkTAW.com. 2005. "Getting Back To Work: A Personal Productivity Toolkit." *http://www.marktaw.com/blog/GettingBackToWork.html* (enter a task for yourself to do next, and keep track with a web app of how many times you have followed through).

Pre-Delete Cruft

Cruft is clutter that bogs things down and gets in the way of getting things done. Idea clutter is mostly stuff that we could have gotten rid of to begin with. When you initiate an activity, determine a kill date for it at the same time.

Computer desktops overflow with icons. Inboxes are filled with ancient email. *Real* desktops overflow with paper: mail, magazines, printouts, notebooks filled with old notes and sums, waiting to be integrated someday (when we have the time) into some master Tower of Babel, stepping us into the stars.

Face it: it's mostly junk, even when we've tried to weed it out along the way. We imagine that we'll use it, and if we think we'll need even a small fraction of it one day, we think we'd better keep it. Some of us are deeply attached to old brilliance and are convinced that our mountains of ideas will be reviewed, collected, prioritized, turned into plans, and converted into fruitful action somehow. Or we worry that *at some point* we're going to *need* one of those little notes, and we're going to be *sorry* that we don't have it. Or perhaps we're worried that we're going to have good ideas for only a limited time, so we start to squirrel them away and hoard them. We spend so much time hoarding them—stacking them, sorting them, working around them, feeling bad about them—that we don't get to implement any of them.

Whether you're attached to your ideas or you're simply having problems with your clutter (a.k.a. *cruft*), here's a little trick that will quickly wipe out most of your future clutter. It's called *pre-deleting*, and it's simple. The only hard part is adjusting your mind into the state where you're willing to do it.

"But I don't want to destroy anything, ever!" Don't worry, we'll address that later in this hack. "But I've got a computer, and it can remember things forever!" We'll talk about that later also. "But it's got terabytes of—" Yes, yes, I know. We'll talk *later*.

In Action

Every time you create or receive something, decide right then how long you're likely to need it. If you're working on something for a couple of weeks, give it a *kill date* two months away. Give it as long a life as it might possibly need, but *pick a date*. Further, record the information in a regular way.

For example, if you write the year that it's safe to destroy a particular piece of paper on the top-right corner in purple ink, *always* do it that way. Make it consistent, and make it easy to see quickly. Then, when you're looking over papers later, you won't have to search all over a page to see whether you can safely throw it away. You can just look in the same spot: "Oh, there, it says I could throw this away two years ago, so *it's trash.*"

With a jaunty grin, crumple up the paper to the astonishment of your friends and co-workers, and casually toss it over your shoulder into the trash bin. Fellow thought-keepers might be shocked—*shocked!*—at your readiness to *destroy data.*

Paper. Whenever you write something down, figure out when it will be safe to destroy it, and mark it accordingly. If you're writing on a page and you know you won't want to keep it a long time, or if you're just using a few sheets of paper to work out some thoughts, put a lightly dashed X in red ink across the front of the page. You should still be able to use the page. This just signals you to "destroy on sight."

Naturally, you'll know not to toss the page the hour or day when you're using it. But when you're done, if you ever see that paper again, you know that you have preauthorized the immediate destruction of the page.

Don't think twice! Just throw it in the trash!

But have *fun* doing so. That gleeful grin on your face can help dispel doubt in your mind. Mark mail as it comes in, magazines, everything that hits your desk. Then respect the kill dates and don't second-guess yourself.

You can even set up files for things that need to be around a bit longer. If today is June 7 and you know you'll need something for two weeks, put it in a file marked "June" and then throw away everything in that file when July comes around. Make a file for July if you feel it will be needed until sometime in July. Just be sure that you aren't moving things forward indefinitely. Be firm and stick with the original kill date.

Computer files. You can mark things similarly when you work on your computer. Create a directory called *tmp*, give yourself permission to throw away anything in it, and then work in it.

Work inside *tmp*, save important files to more permanent directories, and then delete everything in it at the end of the day. Have no doubts: if there were something in it that you intended to keep for longer than a day, *you wouldn't have put it into that directory*. Your project's source code, if you are a programmer, *does not go in tmp*. But keep the text of a temporary file, or a download, or a throwaway project, or anything else with limited utility in your *tmp* directory, and then get rid of it all every day.

If you need to keep something for a couple of days or a week, make folders for the months as you did for paper. In that case, make a *tmpjun* folder for the month of June and delete *tmpjun* when July comes. You can also name the folder *tmp2006* and delete it in 2007.

By now you should have the idea. For any medium, you can find a strategy for communicating to yourself when to stop caring and just throw it out.

When you have those rare things that you really *must* keep for a long time, you'll know it at a glance, because it won't have any red text on it or whatever signal you used for that medium. For those important things, you must do the *opposite* of pre-deletion: set aside a special, protected space for them.

How It Works

The heart of this hack is making it easier to ruthlessly eliminate clutter. Clutter poses several problems.

Clutter demands attention. When you have a lot of clutter, you need to figure out which things you still need and which things you can chuck. If you run into a stack of ideas (in paper or electronic form), you don't necessarily know if you need them. Maybe there's something important in there, right?

So, you pull up a chair, and you go through them one by one. It can be quite an emotional ordeal. If you've pre-deleted, however, there's no heartbreak, no thinking, no consideration. Just tossing.

Clutter makes it harder to find things. If you have clutter, it's obviously harder to find things physically. But this principle applies to computers, too. Even if you have a lot of storage and a wonderful indexing and searching system, it becomes harder and harder to manage and use the material you have as you gain more material. The less you save, the less you have to work through to find what you need, and the faster it goes. You'll find that it's miraculously easier to *do* things, all over the place—at work, on the computer, wherever—because you threw away your junk.

You might lose something, of course, and there might be times that you threw away something you could have used, but you might have lost the same thing in the clutter, too. Given that you've stored truly important things in a safe place, and that almost anything else can be replicated in one way or another, you might find that your newfound agility offers returns in increased concentration that more than compensate.

Clutter fills the future with the past. An interesting thing about records is that we tend to see them again. That's what they're there for, after all. There's something about looking backward, however, which seems to damage the soul.

Reactivating dead thought patterns, over and over again, we can feel old desires like ghosts, moving us this way and that as we put ourselves under their sway again. A 15-year-old boy remembers wanting to become an astronaut. A 23-year-old man laments a lost girlfriend. Even the little things carry ghosts: a shopping list never fulfilled for an old project, a half-finished drawing, a story idea in a line.

If the bad memories nag, happy memories can be even worse. Winning a medal in sixth grade. Old soccer trophies. A special love letter. To be sure, we remember these times with love and fondness, but there is also something *bad* there. There can come to be a strange gnawing feeling and a dissatisfaction with a present that can never live up to the polished memories of old expectations.

Be careful of what you force yourself to remember. Be mindful when sending messages to your future self, because it might not want to be bothered so much.

In Real Life

Pre-delete the trappings of the immediate day. Keep the important mental artifacts as long as you must, but be ready to discard them when you are done. Don't end up saddled with a warehouse full of old ideas and memories so that taking care of them edges out everything else you can do today and tomorrow.

You'll have less stuff pulling at your attention, it'll be easier to find what you want in the present, and you'll set your future free.

—*Lion Kimbro*

CHAPTER THREE

Creativity

Hacks 19–34

Every human achievement is the result of an initial act of creativity. Stone-henge could not have been created without it, nor could the book you are holding. Even some kinds of logic, such as inductive reasoning, require leaps of creativity.

Creativity might appear to be a mystical force, but in fact, it's available to everyone, even people who claim they're not creative. As counterintuitive as it might seem, creativity can be *hacked*. Some of the hacks in this chapter go so far as to try to mechanize creativity. Whether they all succeed is something you'll have to judge yourself, but I hope you'll learn to boost your creativity regardless.

Seed Your Mental Random-Number Generator

Your mind is like your computer's random-number generator: it needs a "seed" from the environment to break out of its routines. You have to put something into it to get something out!

Too often, brainstorming meetings take place in sterile, empty conference rooms with bare walls and nothing to look at anywhere else. They are almost like the industrial clean rooms where microchips are manufactured and not a speck of dust is allowed to gather. Is it any wonder that so many bad ideas come out of these rooms? The truth is that brains need "dust." Brainstorms, like rainstorms, need nuclei around which (b)raindrops can form. If you *start* with no ideas, you will *end* with no ideas.[1]

Think of your mind as a desktop computer faced with the problem of generating a random number out of thin air. Such a computer cannot generate truly random numbers; it can only perform a series of rigid calculations. From the human point of view, randomness enters the computer only when it is programmed to consult its real-time clock for some real-world quantity, such as the number of milliseconds since January 1, 1970. Given this unpredictable input, the PC can then go on to generate output that looks quite random—and even creative. In other words, the computer needs input from an outside source to break out of its rigid patterns.

Humans, too, can become stuck in creative ruts. Everyone has a certain set of interests, ranging from things about which they are mildly curious to those about which they're completely obsessed. Choreographer Twyla Tharp calls this our "creative DNA." Sometimes, this *hardwiring* leads to repetition in our creative output. At that point, we, too, need to seed our mental random-number generators with new data to kick us out of our ruts.

In Action

You can seed your own creative process with almost anything:

- Read a street sign.
- Read a street sign backward.
- Turn on the radio or TV for 10 seconds. (Remember to turn it off! Don't accumulate negative momentum [Hack #65]!)
- Open a book at random.
- Use the random-page feature of the Wikipedia.
- Buy a magazine you would never dream of reading, and read all of it.

Free-associate from the first random stimulus you encounter via one of these methods. What does it make you think of, and how does it relate to your project? You may find po [Hack #21] useful here to join ideas together.

Follow that thought as far as you can, possibly employing techniques such as SCAMPER [Hack #22]. Don't be afraid to be silly. Sometimes a great idea is several silly ideas down the line from the seed. When you reach a dead end, seed again!

If you are having a brainstorming meeting in an empty conference room of the sort already described, bring some books, a magazine, music, pictures, anything that can act as a creative seed. The Oblique Strategies [Hack #23] are explicitly designed with this purpose in mind. If the rut you are in extends not only to your project but also to the rest of your life, try rolling the dice [Hack #49].

In Real Life

While working on my latest board-game design on the bus to work, I passed a sign in an industrial zone that read "American Frame and Alignment." I decided to use the word *frame* as the seed for my brainstorm. One thing led to another, and I soon had a bunch of ideas dealing with *enclosing* games within rule structures, with recursive games and *frames within frames*, like those pictures that show a person holding a picture of herself holding a picture of herself, and so on. This one seed for my mental random-number generator eventually shaped the design of my game system a great deal, and it is now called GameFrame.

It's a sure thing that my "creative DNA" determined the shape of my game, at least partially. For example, I've been fascinated with recursion and self-reference at least since I read *Gödel, Escher, Bach* when I was 14. These powerful inclinations are analogous to the rigid programming of one's PC. However, the street sign bumped me out of my rut. Before I gave my *random-number generator* a *seed* to chew on, the only idea I came up with was that my game might use drinking straws as components!

End Notes

1. Hall, Doug, and David Wecker. 1995. *Jump Start Your Brain*. Werner Books.

See Also

- Random Wikipedia page (*http://en.wikipedia.org/wiki/Special:Random page*)
- Random page from H2G2 (*http://www.bbc.co.uk/dna/h2g2/Random EditedEntry*)
- Other random URLs (*http://randomurl.com*)
- Wikipedia entry for "Random number generator" (*http://en.wikipedia. org/wiki/Random_number_generator*)

HACK #20 Force Your Connections

Use a simple process to generate many complex ideas quickly from a limited pool of simple ideas.

The process of *morphological forced connections* is fairly old; the picture books for children that allow you to combine the head of a giraffe with the body of a hippo and the tail of a fish are one example. The process was formalized by Fritz Zwicky at Caltech in the 1960s[1] and was popularized in 1972 by Don Koberg and Jim Bagnall in their book *The Universal Traveler*.[2]

Most other books that discuss the technique seem to derive their discussion of it from *The Universal Traveler* and even use the same example: creating a new design for a ballpoint pen. We'll take a somewhat different approach.

In Action

The basic process for making forced connections, as outlined by Koberg and Bagnall, is simple and sound:

1. List possible features of the object you are trying to create, one feature per column. For example, the features might include color, size, and shape.

2. In the column under each feature variable, list as many values for that variable as you can. For example, under *color* you might list all the colors of the rainbow, as well as black, white, gold, and silver.

3. Finally, randomly combine the values in your table many times, using one value from each column. To continue our example, you would use one color, one size, and one shape each time.

Technically, steps 1 and 2 are *morphological analysis*, and step 3 is the *morphological forced connections* stage.

The result will be a randomly generated list of possibilities, none of which might be just what you're looking for, but most of which will probably be interesting. Feel free to fine-tune the results. For example, you might not like the suggestion "orange, tetrahedron, a meter on a side," but "orange, tetrahedron, *half* a meter on a side" might hit the spot.

Of course, you can force connections with a pen and paper, as recommended in *The Universal Traveler*, but computers have become widespread since 1972 and you might find them to be a much more efficient tool. To that end, this hack includes a Perl script called *pyro*, which is a somewhat streamlined successor to a HyperCard stack for the Macintosh called Inspirograph that I released in the 1980s.[3]

Inspirograph could generate anything from New England place names (like Lake Nattagoonsucketpocket) to random tabloid headlines. Because the examples I included were humorous, many people who downloaded the stack thought it was only good for a laugh, but it was actually intended for serious design, as is *pyro*.

The Code

Place the following Perl script in a file called *pyro* and make it executable. You also can download the *pyro* script and accompanying *utopia.dat* file from this book's page on O'Reilly's web site (see the Preface for details).

```perl
#!/usr/bin/perl -w
my $infilename = $ARGV[0];
my $basevar = "\@$ARGV[1]\@";

my $pickatrandom = 1;
if (($ARGV[2]) && ($ARGV[2] eq "all"))
{
    $pickatrandom = 0;
}

# Seed the output file
my $outfilename = "/tmp/pyro.txt";
open(OUTFILE, "> $outfilename")
    or die "Couldn't open $outfilename for writing: $!\n";
print OUTFILE "$basevar\n";
close(OUTFILE);

local $/;
undef $/;

open(INFILE, "< $infilename")
    or die "Couldn't open $infilename for reading: $!\n";
$infilecontents = <INFILE>;
close(INFILE);

open(OUTFILE, "< $outfilename")
    or die "Couldn't open $outfilename for reading: $!\n";
$outfilecontents = <OUTFILE>;
close(OUTFILE);

while ($outfilecontents =~ /\@[A-Za-z0-9]+\@/)
{
    local $/;
    undef $/;

    open(OUTFILE, "< $outfilename")
        or die "Couldn't open $outfilename for reading: $!\n";
    $outfilecontents = <OUTFILE>;
    close(OUTFILE);

    # $baseline is the first line in OUTFILE with a variable.

    if ($outfilecontents =~ /^(.*?)(\@\w+\@)(.*?)$/m)
    {
        $baseline = "$1$2$3";
        $varname=$2;
    }
    chomp $varname;

    if ($infilecontents =~ /\#$varname\n(.*?)\n\n/s)
    {
        $varblock = "$1\n";
    }
```

```
        else
        {
            die "Did not find variable $varname in $infilename.\n";
        }

        @varblockarr = split (/\n/, $varblock);
        @outlinesarr = ();

        if ($pickatrandom)
        {
            # Generate a random string from the input elements
            $randline = $varblockarr [ rand @varblockarr ];
            $curline = $baseline;
            chomp ($curline);
            chomp ($randline);
            $curline =~ s/$varname/$randline/;
            push (@outlinesarr, $curline);
        }
        else
        {
            # Generate all possible combinations of the input elements
            foreach $varline (@varblockarr)
            {
                $curline = $baseline;
                chomp ($curline);
                chomp ($varline);
                $curline =~ s/$varname/$varline/;
                push (@outlinesarr, $curline);
            }
        }

        $outlines = join ("\n", @outlinesarr);
        $outfilecontents =~ s/\Q$baseline\E/$outlines/s
            or die "baseline not found.\n";
        $outfilecontents =~ s/\n\n/\n/mg;
        open(OUTFILE, "> $outfilename")
            or die "Couldn't open $outfilename for writing: $!\n";
        print OUTFILE $outfilecontents;
        close(OUTFILE);
    }

    if ($pickatrandom)
    {
        print $outfilecontents;
    }
```

Running the Hack

Here are the first 20 lines of a datafile (called *utopia.dat*) for the *pyro* script.
You can use this file to generate interesting settings for fantasy and science
fiction stories. The values for the variables were culled from two reference
works on speculative fiction.[4],[5]

In case you don't have a computer running Perl handy, or if you'd simply rather not get into running code, I'll explain an alternate way to do the same thing with dice later.

```
#@place@
@type@ made of @material@ @location@, inhabited by @inhabitants@, @govt@,
@special@

#@type@
a cavern
a city
a country
a forest
a jungle
a mountain
a planet
a sealed habitat
a village
an island

#@material@
a superstrong material
crystal
flesh
gold
```

See the "How to Run the Programming Hacks" section of the Preface if you need general instructions on running Perl scripts.

If you have Perl installed on your system, save the *pyro* script and the *utopia.dat* file in the same directory, and then run *pyro* by typing the following command within that directory:

```
perl pyro utopia.dat place
```

If you're on a Linux or Unix system, you might also be able to use the following shortcut:

```
./pyro utopia.dat place
```

Each variable in any *pyro* datafile you create should be wrapped in two @ signs, as in this example. To assign a set of possible values to a variable, place the variable on a line by itself preceded by a hash mark (#).

Each successive line should contain one possible value. Separate variable/value sections with blank lines, and leave two or more blank lines at the end of the file. You can create variables whose values contain other variables, as shown with the place variable in this example.

In Real Life

Table 3-1 shows the data from *utopia.dat* in tabular form. Each column of text contains the 10 possible values for one variable, whose name is at the head of that column. There are six columns of 10 items, so you can generate 10^6 or a million possible fantasy places from this data.

Table 3-1. Data to generate fantasy places

No.	Type	Material	Location	Inhabitants	Govt.	Special
0	A cavern	A super-strong material	In an unknown place	Aliens	A socialist utopia	Exceedingly war-like
1	A city	Crystal	In another universe	Apes	An anarchy	Where cannibalism is practiced
2	A country	Flesh	In space	Completely normal people	Ruled by a corporation	Where everyone is happy
3	A forest	Gold	In the desert	Fairies	Ruled by a council of many species	Where everyone is insane
4	A jungle	Paper	In the dream world	Fish	Ruled by a god	Where learning is exalted
5	A mountain	Porcelain	In the fourth dimension	Ghosts	Ruled by a hereditary monarch	Where telepathy is common
6	A planet	Rubber	In the polar regions	Giants	Ruled by a magical elite	Which is boundless
7	A sealed habitat	Stone	In the sky	Insects	Ruled by ancient ritual	Which is microscopic
8	A village	Water	Under the Earth	Intelligent plants	Ruled by computer	Which is sacred
9	An island	Wood	Under the sea	Robots	Ruled by women	Whose location changes

If you have a 10-sided die (or, even better, six of them), or if you have some other way to generate random digits, you can create fantasy places using Table 3-1, generating digits sequentially and selecting one item from each column. For example, Utopia #895779 is "a village made of wood in the fourth dimension, inhabited by insects, ruled by an ancient ritual, whose location changes."

The *pyro* script will do the same thing faster. It takes two mandatory arguments and a third optional one:

- The first argument is the name of the datafile to use, such as *utopia.dat*.
- The second argument is the name of the variable from the datafile that you want to expand, such as place. (Do not wrap the variable name in @ signs to match the datafile; *pyro* will do that for you.)
- The third argument, which is optional, is the word all. If this option is used, *pyro* will generate all possible combinations of the elements in the datafile, instead of one random element.

Generating all possible combinations with the all argument might take some time.

Thus, the following command will generate all one million possible fantasy places and leave the results in the file */tmp/pyro.txt*:

```
./pyro utopia.dat place all
```

But this command will simply generate one fantasy place:

```
./pyro utopia.dat place
```

Here's a slightly edited console log of running the previous command:

```
$ ./pyro utopia.dat place
a mountain made of rubber in the sky, inhabited by fish, ruled by a
hereditary monarch, where learning is exalted
$ ./pyro utopia.dat place
an island made of crystal in the fourth dimension, inhabited by aliens, an
anarchy, which is microscopic
$ ./pyro utopia.dat place
a sealed habitat made of paper in the dream world, inhabited by giants, a
socialist utopia, where learning is exalted
$ ./pyro utopia.dat place
an island made of flesh in the fourth dimension, inhabited by fairies, an
anarchy, which is boundless
$ ./pyro utopia.dat place
a mountain made of a superstrong material in the dream world, inhabited by
apes, an anarchy, where cannibalism is practiced
$ ./pyro utopia.dat place
a cavern made of porcelain in the polar regions, inhabited by intelligent
plants, ruled by a council of many species, which is microscopic
$ ./pyro utopia.dat place
a sealed habitat made of stone in an unknown place, inhabited by giants,
ruled by a magical elite, where everyone is insane
```

```
$ ./pyro utopia.dat place
a sealed habitat made of flesh in the desert, inhabited by giants, ruled by
a corporation, where cannibalism is practiced
$ ./pyro utopia.dat place
an island made of paper under the sea, inhabited by fairies, an anarchy,
which is boundless
$ ./pyro utopia.dat place
a forest made of flesh in an unknown place, inhabited by fairies, ruled by
computer, exceedingly warlike
```

While *pyro* is useful as it stands, it's certainly possible to improve it. More robust error checking could be added (such as checks for circular variable definitions, and lack of a newline at the end of the file or between entries), and it might be nice if the number of random strings to be generated were an alternate value for the third command-line option. It's open source, so hack away!

End Notes

1. Ritchey, Tom. 2002. "General Morphological Analysis: A general method for non-quantified modeling." *http://www.swemorph.com/ma.html*.

2. Koberg, Don, and Jim Bagnall. 1976. *The Universal Traveler: A Soft-Systems Guide to Creativity, Problem-Solving, & the Process of Reaching Goals*. William Kaufmann, Inc.

3. The original Inspirograph. *http://ron.ludism.org/cloudbusters/inspirograph-10.hqx*.

4. Manguel, Alberto, and Gianni Guadalupi. 1980. *The Dictionary of Imaginary Places*. Macmillan Publishing Co., Inc.

5. Stableford, Brian. 1999. *The Dictionary of Science Fiction Places*. The Wonderland Press.

See Also

- Raymond Queneau of the French literary group, the Oulipo [Hack #24], used morphological forced connections in his 1960 book *Cent Mille Milliards de Poèmes*, or *One Hundred Thousand Billion Poems*, a flip book with 10 possible strips for each of the 14 lines of a sonnet, hence 10^{14} or 100,000,000,000,000 sonnets. As usual, computers make the whole thing easier. Here's a decent web version in English: *http://www.bevrowe.info/Poems/QueneauRandom.htm*.

- The Oulipo member Harry Mathews also devised a forced-connections procedure dubbed Mathews's Algorithm, which is neither random nor exhaustive. You can read about it and experiment with it online: *http://bumppo.hartwick.edu/Oulipo/Mathews.php*.

Contemplate Po

Use a new word to examine seemingly impossible alternatives, juxtapose random ideas, and challenge stale concepts.

Creativity expert Edward de Bono invented the word *po* to shake up people's thoughts. He listed several etymologies for it. One is that it can be seen as "arising from such words as hy*po*thesis, sup*po*se, *po*ssible, and even *po*etry"; another is that it stands for *provocative operation*, a kind of mental hack to get ideas "unstuck" and move them forward.[1]

Wherever the word comes from, po is a great tool for playing with ideas and seeing the potential surrounding them, without getting too caught up in the details.

In Action

Provocative operations with po come in three basic kinds, which de Bono calls PO-1, PO-2, and PO-3.[2] Each is useful to provoke certain kinds of thinking and move a creative situation forward in a different way.

PO-1. PO-1 means using po to protect a "bad" idea from premature judgment so that it can be used as a stepping-stone to genuinely good ideas. For example, if you are considering solutions to the problems that the U.S. space program has suffered, you might say to yourself, "Po the space shuttle should be blown into a million pieces." Normally, blowing up the space shuttle would be a bad idea, but po "protects" it so that it can lead to potentially good ideas.

You don't think about and judge that specific idea, but focus instead on ideas that come from it. In this example, one idea might be a group of smaller, modular vehicles holding only one person that assemble into a larger station when in space, and then break apart again for reentry. Not only might this be cheaper and easier to produce, but also, in case of a disaster, only one person would be killed instead of the whole crew.

PO-2. PO-2 means using po to juxtapose ideas randomly to help you seed your mental random-number generator [Hack #19]. Suppose you are trying to develop a new idea for a game. You might provoke yourself with the phrase "game po eyeglasses," which might then lead to the following ideas:

- An inexpensive video game system built into a set of goggles that project the graphics directly onto the retina
- A scavenger hunt for charity, with prizes for the team that collects the largest number of used eyeglasses to be recycled in developing countries

- A dexterity game in which people who wear eyeglasses are handicapped by having to take them off and people who don't are handicapped by having to put them on

Probably not all of these ideas would be useful to you, but some might, and if they aren't, you can replace *eyeglasses* with some other word.

PO-3. PO-3 means using po as a creative challenge for change. It is used to set aside an existing idea, or parts of one, without actually rejecting it.

For example, consider the statement made in a meeting you are attending: "Our group should meet every week, as we always have." You might ask the following questions:

- "Po *every*: why not alternate weeks?"
- "Po *week*: why not every five days, or every month?"
- "Po *group*: why doesn't our group break into multiple smaller, overlapping special interest groups that can meet when they want?"

How It Works

The subtitle of de Bono's book *Po* is *Beyond Yes and No*. According to de Bono, the problem with much of human thought is that we prematurely accept or reject ideas with *yes* or *no*, and once we accept one, we cling to it far longer than is useful.

The word *po* is intended to provide a kind of breathing space outside of what de Bono calls the "YES/NO system," where we can pause and coolly consider all the possibilities. Metaphorically speaking, instead of automatically turning to *yes* or *no* when we see an idea, we can jump *up* and look at them both from a higher viewpoint.

Douglas Hofstadter has another word for this: he calls it *jootsing*, for *jumping out of the system*.

In Real Life

Po is a terrific tool for isolated brainstorming scenarios, but it can be highly useful in the heat of everyday life as well. For example, if someone cuts you off in a car, your natural reaction might be to get hot and lean on the horn, cussing loudly. If instead you mutter "po" to yourself, you might remind yourself to invoke what general semantics calls a *semantic pause* or the *wedge of awareness*:[3] a pause to reflect on what you are doing before you do it. You can then ask yourself whether it's really worth upsetting yourself over the incident.

Similarly, if your spouse knows the word *po*, one of you can say it to remind yourselves that an argument need not have a winner and a loser, but that both parties can retain their different viewpoints. By saying "Po!" you're saying, "Let's agree to disagree." You can also use it to reach a reasoned decision together. When you're tempted to run to opposite sides and dig in your heels, you can both step up onto po and look at the situation from the same calm spot.

End Notes

1. de Bono, Edward. 1999. *Six Thinking Hats*. Back Bay Books.
2. de Bono, Edward. 1972. *Po: Beyond Yes and No*. Penguin Books.
3. Dawes, Milton. "The Wedge of Consciousness: A Self-Monitoring Device." *http://miltondawes.com/md_wedge.html* (well worth reading if you find this hack useful).

See Also

- The Po Machine (*http://www.davidaspitzley.org/PoMachine/Index.asp*) conjoins words randomly with PO-2 and collects visitors' connections between them.

HACK #22 Scamper for Ideas

SCAMPER is a mnemonic acronym for a set of basic operations that you can apply to old ideas to extend them in new directions.

The SCAMPER (Substitute, Combine, Adapt, Modify, Put to another use, Eliminate, Reverse) technique is a highly portable creativity toolbox. It was developed by Bob Eberle as a unified set of brainstorming tools with a simple mnemonic and was first published by Michael Michalko in his book *Thinkertoys*.[1]

You can't get data off a computer with a blank hard drive, and you can't get creative ideas out of your brain without having put something into it first. SCAMPER is a *structured* way of seeding your brain's random-number generator [Hack #19] to produce creative output when you feel uninspired.

In Action

Use SCAMPER as a checklist. Choose an object to think about creatively, such as a drinking cup, and then run down the items in the mnemonic checklist (Substitute, Combine, Adapt, and so on), asking yourself the question associated with each one in Table 3-2, which explains the basic structured brainstorming techniques of SCAMPER. The word *target* refers to the object you're thinking about, such as the drinking cup in the previous example.

Table 3-2. SCAMPER mnemonic checklist

Mnemonic	Key	Questions to ask
S	Substitute	How can you *substitute* something else *for* the target or *within* the target?
C	Combine	How can you *combine* something else with the target to produce something new?
A	Adapt	What techniques, mechanisms, or components can you *adapt* to the target from elsewhere?
M	Modify/magnify	How can you *modify* the target in a useful way? What aspects of the target should you *magnify* or increase?
P	Put to another use	How can you *put* the target *to other uses*?
E	Eliminate/minimize	What should be *eliminated* from the target, or *minimized* in it?
R	Reverse/rearrange	What is the opposite (*reverse*) of the target? What *rearrangements* are possible within the target or by using it?

For example, if the item is Substitute and you're working with the drinking cup, ask yourself, "How can I substitute something else for the cup? How can I substitute something else *within* the cup?" You might come up with the idea of a new kind of bottle to drink from instead of a cup (here in hydrated Seattle, everyone I know drinks from Nalgene bottles and the like), or you might develop the idea of substituting a better material for the one the cup is made of. What about an indestructible titanium cup? The only cup you'll ever need!

> You can generalize the technique used to create SCAMPER itself and assemble your own mental toolbox **[Hack #75]** that goes beyond brainstorming, including hacks for memory, mental math, critical thinking, and so on.

How It Works

The SCAMPER technique is basically a cross between seeding your mental random-number generator **[Hack #19]** and using mnemonics (your dear, dear friend) just as you can use them to remember 10 things to bring when you leave the house **[Hack #1]**. In this case, however, the seeding is done in a very structured way, with "known fruitful" mental seeds, and the mnemonics are used to remember to carry mental tools with you, instead of physical ones.

In Real Life

I recently applied SCAMPER to the traditional card game Rummy (*http://pagat.com/rummy/rummy.html*) to develop the basis for a new card game, Scrummy.

In case you haven't played Rummy, it's a card game based on collecting *sets* (cards with the same number, such as 5, 5, and 5) and *runs* (cards in sequence, such as 5, 6, 7, and 8). These cards are laid on the table in groups called *melds*; players can also add to a meld on the table by *laying off* their own cards in front of themselves. A Rummy game ends when a player *goes out* by playing her last cards.

Here are some ideas that came to me by applying SCAMPER operations to the basic rules of Rummy. Note that I didn't use the SCAMPER tools in strict order. You don't need to, either. Also, I might not use all the ideas I generated in the final game. SCAMPER is best used for brainstorming new ideas; editing the possibilities is a job for later development.

Rearrange

> Sets and runs can be rearranged to make new kinds of melds, such as the digits of π (3, 14, 15, 9...) or even numbers only (2, 4, 6, 8 ...).

Put to another use

> Use Scrummy to select a winner for a prize, such as getting a back rub, not having to cook dinner, or picking the next game to play.

Substitute

> Use a deck of cards other than the standard deck to play, such as Sticheln, a six-suited German deck that has three suits with cards from 0–18 and three suits that range from 0–20.

Combine

> Combine card play with dice play. For example, if I decided to use the Sticheln cards, I could specify a 20-sided die (d20) and link rolling a 1–20 on the die with the numbers on the cards from 0 to 20 to make certain special actions possible.

Adapt

> There is already a game system that combines dice with special cards, and it's called the Dice Deck. Can you play a kind of rummy with the Dice Deck? Yes: Dice Deck Rummy. If you roll a 7 during Dice Deck Rummy, you can lay down a set of sevens (such as 7, 7, 7) or a run that begins or ends with 7 (such as 4, 5, 6, or 8, 9, 10). The single 20-sided die should work well if I adapt this.

Eliminate/minimize

> Eliminate players' melding or laying off in front of themselves: each player must lay cards directly in front of another player.

Reverse

> Points in Scrummy are negative rather than positive: you try to avoid them rather than score them, as you do in Rummy. Other players receive negative points when you lay cards in front of them.

Modify

> Modify the concept of "going out." Replace it with "maxing out." No one ends the game by going out when his hand is empty; he just replenishes his hand. Maxing out happens when a player scores -100 points; she is then out of the game.

Magnify

> Magnify the "petty diplomacy" common in some games. Through the rules, encourage players to gang up and knock one another out.

And so, by applying SCAMPER to an old game, I quickly developed the outline of a new game. What new places can you SCAMPER to?

End Notes

1. Michalko, Michael. 1991. *Thinkertoys*. Ten Speed Press.

HACK   Deck Yourself Out

#23   Creativity decks are sets of suggestive aphorisms for getting creatively unstuck—for example, unblocking your writer's block. They come in several forms: as decks of cards, as PC and PDA applications, and even as scripts that you can consult via the Web, wherever you are.

Many hacks in this book use some form of randomized input to stimulate creativity or to break deadlocked decisions, such as seeding your mental random-number generator [Hack #19], rolling the dice [Hack #49], not overthinking it [Hack #48], and forcing connections [Hack #20]. This should not surprise you. Human use of random stimuli is ancient, ranging from staring at the clouds, to reading entrails, to poring over Tarot cards. The human brain is well adapted to finding patterns in a wide range of stimuli, so it's smart to use this natural functionality to focus conscious thought and capture less-conscious thoughts.

In recent years, a new genre of random stimulus has been developed that I'll call the *creativity deck*. Designers of creativity decks aim to fill them with ideas and strategies that are good in themselves and don't require so much dreamy dissociation to work—as cloud gazing does, for example. These decks have become enormously popular, so there are now many to choose from. And they're available in many different formats, so you can select a deck and format to suit your tastes and needs.

In Action

This hack will examine three of what I consider the most important and popular creativity decks currently available: the Oblique Strategies, the Creative Whack Pack, and the Observation Deck.

The Oblique Strategies. The Oblique Strategies are a creativity deck developed in 1975 by British painter Peter Schmidt and musician Brian Eno (noted for his mind music [Hack #27]). They have since gone through several editions.

The strategies were originally intended for musicians and other artists. Each card contains a strategy for solving a problem in the studio or another work environment. The cards are intended to remind the user of ideas that might not seem obvious under pressure. All of the strategies were originally important working principles for someone: initially Eno and Schmidt, and later their friends and colleagues such as Stewart Brand of the *Whole Earth Catalog*.

Although no magical powers are claimed for the deck, an Oblique Strategies consultation has a subtle, mysterious, gnomic feel to it, somewhat like the I Ching. Often, the user has to tap intuition to interpret a card, which adds to the divinatory feeling. Here is the text from some typical cards:

- Ask people to work against their better judgment
- Which parts can be grouped?
- Take away the elements in order of apparent non-importance
- Go to an extreme, move back to a more comfortable place
- Water

Over the years, the original Oblique Strategies have been made available in multiple editions and in a variety of formats. For example, they were recently released as a widget for Mac OS X.[1] Figure 3-1 shows what the widget looks like before and after turning over the virtual card.

The strategies themselves have also become applicable to more situations than just music and painting, although vestiges of the original purposes remain. If you are a smart, creative person who prefers elegant ambiguity to having everything spelled out, the Oblique Strategies deck might be for you.

You can find Oblique Strategies applications for numerous platforms, including the Palm, at the official Oblique Strategies web site.[2] As I mentioned, you can also consult the strategies online.[3]

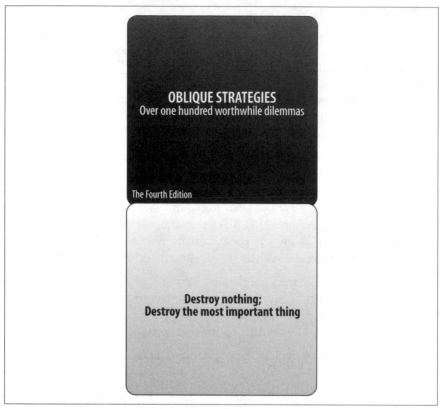

Figure 3-1. The Oblique Strategies widget for Mac OS X

The Creative Whack Pack. The popular Creative Whack Pack,[4] developed by
Roger von Oech in 1992, is similar to the Oblique Strategies deck in concept, but is positioned more for businesspeople than artists.

The deck has 64 cards with text and graphics taken primarily from von
Oech's book, *A Whack on the Side of the Head.* The cards are split into four
suits of 16 cards each: Explorer (blue), Artist (orange), Judge (green), and
Warrior (red). These suits represent von Oech's view of the main roles in
creative thought, similar to Edward de Bono's *Six Thinking Hats.*[5]

You can obtain the Creative Whack Pack and related materials from the
author's web site,[6] where you'll also find an online version of the deck that
you can try. Figure 3-2 shows a sample page with the text and image of a
typical Judge card from the Creative Whack Pack.

The deck comes with numerous exercises other than just drawing a card
and following instructions. For example, the "Three Day Agenda" exercise
instructs you to pick five cards for actions you'd like to take in the next three
days, returning those cards to the deck as you complete them.

A leading business school did a study that showed that its graduates did well at first, but in ten years, they were overtaken by a more streetwise, pragmatic group. The reason according to the professor who ran the study: "We taught them how to solve problems, not recognize opportunities." Where do you hear opportunity knocking? How can you answer it? What opportunities are present in your current problem?

Figure 3-2. An online sample card from the Creative Whack Pack

The Observation Deck. The Observation Deck[7] is a creativity deck aimed especially at writers with writer's block. It was developed in 1998 by Naomi Epel, an author whose day job is literary escort (making sure that authors on book tours get fed, arrive at their venues on time, have necessary supplies, and so on). In the course of escorting authors around San Francisco, she grilled them on their own mind performance hacks, which she then distilled into a deck of 50 cards and an accompanying 160-page book.

When stuck, a writer draws a card from the deck and looks up the corresponding chapter in the book. The titles are brief and evocative. Here are a few:

- "Zoom In and Out"
- "Rearrange"
- "Feed Your Senses"
- "Switch Instruments"

The chapters are too long to quote in full, but by way of example, the "Switch Instruments" chapter suggests trying different ways of writing, such as dictation, handwriting, or writing in another language. It backs up these suggestions with anecdotes about authors such as Spalding Gray and T.S. Eliot.

Unfortunately, there is no online version of the Observation Deck, as there is with the Oblique Strategies and Creative Whack Pack decks, but at the time of this writing, the Observation Deck is available from most major brick-and-mortar bookstores and online booksellers.

In Real Life

To get started on writing the current hack, I pulled a card from the Observation Deck. It was "Ask a Question." The inspiration for the card came to Epel from columnist Jon Carroll, who quoted Rudyard Kipling when asked where he got his material:

> I have six humble serving men
> They taught me all I knew
> Their names are what
> and where and when
> and why and how and who.

Epel describes her interaction with Carroll in the book that accompanies the Observation Deck, and suggests there that the reader create yet *another* deck of cards. She writes:

> Create a deck of question cards to use whenever you feel stuck. Print one question on each of seven blank cards—you should have six serving men cards and one that reads "What if?"

Epel's idea is that the reader should draw a question card at random and then use that question to go deeper into his material. When I started writing the current hack, I made my own "serving men" deck. I drew the "Who?" card first, so I started by learning more about the creators of the various creativity decks I was going to describe, such as Brian Eno and Peter Schmidt, Roger von Oech, and Naomi Epel herself. I didn't use all the material I gathered, but it was a great jump-start.

Here's a somewhat less artificial example of using a creativity deck. On one recent technical writing job, I was working with a particularly intransigent literate programming tool—and some particularly intransigent upper management. I consulted the Oblique Strategies. I think the strategy I drew was "Disciplined self-indulgence," which led me to ask myself which parts of my job I enjoyed the most and whether I could use them to solve the problem. One of the aspects of my job that I enjoyed most was scripting the documentation build process, usually with Perl, so I came up with the idea to post-process the troublesome documentation with a series of Perl scripts, which I quickly wrote.

When I showed the final product to my manager, she was amazed and all but called me a magician. She asked me how I had gotten the idea for the solution, so I showed her the online Oblique Strategies application I had consulted. She was crestfallen. Suddenly, I was no longer a magician, but a mountebank.

In this case, the Oblique Strategies certainly helped me solve the problem, but I found a possible pitfall with creativity decks: people who don't understand them may believe that the decks are doing your thinking for you.

End Notes

1. The Mac OS X widget for the Oblique Strategies is available at *http://www.apple.com/downloads/dashboard/reference/oblique.html*.

2. The Oblique Strategies web site is located at *http://www.rtqe.net/Oblique Strategies*.

3. You can find Oblique Strategies online at *http://stoney.sb.org/eno/oblique.html*.

4. von Oech, Roger. 1992. *Creative Whack Pack*. U.S. Games Systems, Inc.

5. de Bono, Edward. 1999. *Six Thinking Hats*. Back Bay Books.

6. Buy the Creative Whack Pack, find related materials, and view sample cards at *http://www.creativethink.com*.

7. Epel, Naomi. 1998. *The Observation Deck*. Chronicle Books.

See Also

- The Idea Generation Methods site (*http://www.ideagenerationmethods.com/methods/card-decks.html*) lists a number of other creativity decks.

- You can buy an official Oblique Strategies deck of the latest edition online (*http://www.enoshop.co.uk*) for 30 British pounds. If you'd like a real deck of Oblique Strategies cards that you can hold in your hand, but you don't want to pay more than $50 plus shipping from the UK, you can obtain a printable PDF deck from *http://www.stretta.com/~matthew/resources/oblique/index.html*. (I have also created a printable deck based on the combined list of strategies from the GPLed Palm application at the official Oblique Strategies site and may make it available someday.)

- A GreaseMonkey script (*http://masculinehygiene.com/sburke/Oblique_Strategies.user.js*) for Oblique Strategies is available for download at *http://dunck.us/collab/GreaseMonkeyUserScriptsGeneric*.

Constrain Yourself

HACK #24

Rigidly constrain your creative work to find the potentially fascinating emergent effects that happen when you have to overcome artificial obstacles.

Because the word *constraint* has such negative connotations in our culture, and because this book is about freeing your mind, you might be wondering if I'm about to attempt an Orwellian *war is peace, freedom is slavery, ignorance is strength* reversal. Examples of paradoxically productive, even freeing constraint familiar to people in our culture aren't hard to come by, however: for example, it is actually easier to write good poetry with rhyme and meter than good free verse, because rhyme and meter *force* your pen to interesting places it would never normally visit—a potent mind performance hack bequeathed to us by the ancients!

Some French artists known collectively as the Oulipo (*Ouvroir de Littérature Potentielle*, or Workshop for Potential Literature) have brought the theory and practice of creative constraints to a refinement never before experienced. Unconstrained writing, such as free verse or ordinary prose, is of no interest to the Oulipo. *Rules* and *games* as applied to literature are the sole reason for being of the group, which was founded in 1960. Even rhyme and meter are considered insignificant as constraints by these pioneers, who prefer to write enormous palindromes, novels without the letter *E*, and farces whose structure is determined by the mathematical principles of combinatorics.

There are many media to which constraints can be applied besides literature, so you can still use this hack even if you are not a writer. Oulipian offshoots for different art forms include the Oubapo (for *bandes dessinées*, or comic strips), Oucinepo (cinema), Oucuipo (cuisine), Ouhistpo (history), Oumupo (music), Oupeinpo (*peinture*, or painting), and Ouphopo (photography); experiments have even been performed in Oulipian computer programming and mathematics.

Right about now, you might be wondering where the *po* or *potential* in *potential literature* comes from. Essentially, the Oulipo party line is that the artistic potential of a work is generated by the constraints placed upon it. An analogy I like to use is squeezing a tube of toothpaste: the constraints on the tube are your fingers, which generate physical pressure, analogous to potential, which in turn produces—squirt!—toothpaste, or inspiration. Due diligence alone will produce the creative work itself; altogether, I like to call this the "Muse helps those who help themselves" model of creativity or, more irreverently, "praise the Muse and pass the inspiration!"

In Action

Likely, you have a creative project to which you would like to apply this method. If so, start thinking about constraints that might apply to it. Find one or two that suit you from the following list or from one of the sources listed in this hack's "End Notes" section. Exploit it to the max!

A few sample constraints can be illustrated with text and music.[1,2]

Text. Usually, you'll be doing some form of *writing*; fortunately, this is the most heavily explored medium. Here are a few sample constraints for text:

Snowballs
> Poems in which each successive word is one letter longer (or each word one syllable longer, or each sentence one word longer, or each paragraph one sentence longer); sometimes the process reverses midway, and this is called a *melting* snowball.

Exercises in style
> Tell the same story over and over from multiple perspectives (similar to changing your worldview [Hack #31]).

Record setting
> Taking an ordinary constraint of a text and setting a record for the longest text using that constraint, or for the text that uses that constraint the most, and so on (for example, the palindrome created by one prominent Oulipian was more than 5,000 letters long).

S+7
> A famous Oulipian technique in which every noun in a piece of writing is replaced by the seventh noun following it in a specified dictionary; this tends to produce eerie nonsense, which might be what you want, or might just be useful as a way to seed your creative random-number generator [Hack #19].

Music. *Music* can also be created with Oulipian methods:

Alphabetical constraints
> Although the Oulipo tend to abhor randomness, you can use musician Brian Eno's Oblique Strategies [Hack #23] nonrandomly by ordering the deck alphabetically and (for example) choosing every 12th strategy card to create a complex constraint.

Notation
> New musical notations can be developed, as in Andrew Hugill's choral piece *Révélations et Diversités*, whose score is a kind of giant chessboard across which each singer "moves" like a game pawn.

Unplugging notes

From an existing musical piece based on some predetermined criterion. Christopher Hobbs's systematic excision of the musical notes from classical pieces corresponding to the letters in the name BACH has reportedly produced some interesting results.

Arrhythmic and microtonal music

Music without apparent rhythm, or using pitches that are minutely differentiated as compared to the traditional Western scale, are also areas that have been explored by Oumupo groups, and which you can explore, too.

Limits

Limits can be broken (see "Record setting," described earlier). For example, the score of the Funeral March for the Burial of a Great Deaf Man (Alphonse Allais, 1897) is entirely blank, anticipating John Cage's 4'33" by 50 years (ironically so, since the Cage estate sued composer Mike Batt for recording silence on an album, and then settled for a six-figure sum).[3]

In Real Life

Being a great fan of the Oulipo, I have used their techniques many times; for example, in the 1990s, there was an explosion of interest in the artistic form of the Glass Bead Game as imagined by novelist Hermann Hesse. You might like to explore thinking analogically [Hack #25]; my own experiments in formal constraint, such as recursive use of the Old Norse poetic form called the kenning, were highly influenced by the Oulipo.

Generally speaking, the influence of the Oulipo has been historically somewhat narrow. Even so, many people have used Oulipian techniques without knowing it. Oulipians jokingly consider such people to be unconscious plagiarists and call their forebears (Lewis Carroll, for example) *anticipatory plagiarists*. Read the Oulipo, then, for examples of the highest development of the Oulipian art. Graeco-Latin bi-squares toured by a chessboard knight are simply not used every day to plot novels! Examples of "true" Oulipian work by writers in English are therefore hard to cite. Still, you find Oulipian constraints where you least expect them....

Perhaps some of the works you read every day incorporate constraints without your knowing it; there are two schools of thought in the Oulipo, one of which says a work's constraints should be made utterly explicit, the other that they should be hidden. E-Prime [Hack #52] is used as a hidden constraint in some of the psychological texts of Albert Ellis [Hack #57]. Reversed is the situation of the French novel *La Disparition* (as well as its English translation,

A Void, by Gilbert Adair), in which the disappearance of the letter E from the world (and the text) is made the central plot element and a highly explicit constraint. Explicitness of constraint, or the lack thereof, can itself be a constraint on a work, of course. Catalog the number of constraints you can exploit such as explicitness, exercises in style, record setting, E-Prime, and so on; then try a few, and you'll see that these seeming shackles are really keys to greater creativity.

> Even the body of this hack is a real-life example of incorporating two stringent, semi-hidden textual constraints into a piece. (1) If you take the first letter of every sentence in the body of the text, they form the sentence *BE SURE TO READ LIFE A USERS MANUAL BY GEORGES PEREC*; each word corresponds roughly to one paragraph. (2) Neither the central Oulipian author Georges Perec nor his masterpiece *Life: a User's Manual* is ever mentioned explicitly in the hack, although both are alluded to several times. This mirrors the disappearance of the central letter *E* in *A Void*, which Perec also wrote.

End Notes

1. Motte, Warren F. 1986. *Oulipo: A Primer of Potential Literature*. University of Nebraska Press.

2. Mathews, Harry, and Alastair Brotchie. 1998. *Oulipo Compendium*. Atlas Press.

3. BBC News. 2002. "Silent music dispute resolved." *http://news.bbc.co.uk/ 1/hi/entertainment/music/2276621.stm*.

Think Analogically
HACK #25
Use analogies to solve problems and extend old ideas in new directions.

In "Enjoy Good, Clean Memetic Sex" **[Hack #26]**, I compare thinking to sex and explore the consequences of that analogy. However, I don't explore the process of comparison itself and how to elaborate the analogy derived. Making analogies is an excellent thinking hack, and the techniques for doing so are worth exploration.

Recent studies indicate that much creative thought is the result of *cognitive blending*—mapping the elements of one idea onto another—a primary form of which is analogy.[1] Thinking analogically can help you to create more prolifically,[2] and recent advances in formalization of analogical thought mean that you can understand the richness available ever more rigorously.[3]

In Action

Although there are many ways to approach the process of creating and exploring analogies, this hack focuses on two of them—*tables of correspondences* and *kennings*—and the ways in which they're related.

Tables of correspondences. First, consider Table 3-3, which summarizes some of the extended thought/sex analogy.

Table 3-3. Simple comparison of thought and sex

Thought	Sex
Brain	Genitalia
Reading	Insemination
Understanding	Conception
Creation	Offspring

Each item in the first column corresponds to the item in the same row in the second column. For example, *creation* in the Thought column corresponds to *offspring* in the Sex column.

Table 3-3 is also similar to an old concept, the *table of correspondences* used by medieval alchemists and other occultists. Such a table might show that the metal corresponding to the sun is gold, and the corresponding animal is the lion. You are not expected to believe this; think of it as poetry for now.

Kennings. Transhumanist thinker Hans Moravec wrote a book called *Mind Children*[4] in which he made the point that robots, memes, and other human mental creations are our descendants, in a way. Since the phrase *mind children* is synonymous with *thought offspring*, perhaps Moravec was using a metaphor that can be derived from the top and bottom rows of our table:

$$\frac{\text{thought}}{\text{creation}} :: \frac{\text{sex}}{\text{offspring}}$$

Moravec's analogy that human creations are our thought offspring can be derived by starting from *creation*, moving up across the line to *thought*, and then jumping diagonally to *offspring*. In fact, four metaphors can be derived from this analogy:

- Thought = creation sex
- Creation = thought offspring
- Sex = offspring thought
- Offspring = sex creation

Some of these metaphors might not make sense until you ponder them a while. For example, to call sex *offspring thought* is to say that it is a deliberate, creative process that produces offspring. (Rephrasing it as *the thought of offspring* might make this clearer.)

I call each of these two-term metaphors a *kenning*, because they work like that Old Norse poetic form, which called the sea a *ship road* and a sword a *flame of battle*.[5]

Although you might not be an alchemist trying to transmute base metals or a skald reciting an epic poem in old Norway, you might be surprised at how useful tables of correspondences and kennings still are.

For practical examples, one of them worth tens of millions of dollars, see the "In Real Life" section of this hack.

You can use tables of correspondences to generate great gobs of kennings with a simple procedure. Here are the five steps of that procedure, generating the *thought offspring* kenning from Table 3-3 as an example:

1. Select a target term for which you wish to find a kenning, such as *creation*.
2. Select another term in the same column as the target, such as *thought*.
3. Select a term in the same row as the target, such as *offspring*.
4. Put terms 2 and 3 together to make a kenning, such as *thought offspring*.
5. Optionally replace the terms of the kenning with synonyms, such as *mind children*.

Depending on the table of correspondences, you might have to rotate the table 90 degrees mentally so that the rows become columns and the columns rows, before this procedure will work.

A table of correspondences can have an unlimited number of rows and columns. Table 3-4 shows the result of adding a single Metabolism column to the Thought and Sex columns of Table 3-3.

Table 3-4. Table of correspondences with three columns

Thought	Sex	Metabolism
Brain	Genitalia	Digestive tract
Reading	Insemination	Eating
Understanding	Conception	Digestion
Creation	Offspring	???

The thought-as-metabolism meme runs deep in our culture. Consider this Francis Bacon quote from Elizabethan times:

> Some books are to be tasted, others to be swallowed, and some few to be chewed and digested.[6]

Consider also the magazine *Reader's Digest*: predigested, pre-understood material, like the food a mother bird regurgitates for her young. When a metaphor such as *understanding is digestion* runs this deep in a culture, it is like a vein of precious metal: it might be tapped out already, or fantastic riches might be waiting just a little deeper.

The table of correspondences is your jackhammer for mining our culture. The poet John Donne was famous for his metaphysical conceits, elaborate metaphors that ran all the way through some of his poems. A bigger table of correspondences, however, will enable you to create a metaphor of literally book length or longer. Since an analogy is a way of understanding something, a complex analogy will help you understand a complex phenomenon. What you choose to understand, and what you do with that understanding, is up to you.

The main pitfall of this technique is the *forced* analogy. Sometimes a correspondence simply doesn't exist. If you pore over many a quaint and curious volume of forgotten lore, you'll find that traditional tables of correspondences often contain bizarre "knowledge," such as the correspondence of mugwort to quartz, or that of musk perfume to the Power of the Evil Eye. Don't give in to the temptation to fill blank spaces with just anything, or things that have tenuous connections; your analogies will suffer. Also, beware correspondences with cultural overtones that might be false or culturally insensitive, unless you don't mind the risk of offending people with your ideas.

Filling in the blanks. You might have noticed that the bottom-right cell in Table 3-4 is blank. What should go there? I would argue *nutrition* or *energy*, because it is the end product of metabolic processes, just as offspring are the end product of sex and creativity is the end product of thought.

You might prefer to replace *energy* with *excrement* if you have an anal-expulsive personality, or you might come up with an entirely different answer. The point is that a blank space in a table of correspondences is an opportunity to be creative.

In Real Life

One of my most successful recent game designs grew from filling in a blank that I noticed while making a table of correspondences about games. While the card game Hearts is so popular that it has been adapted to many different types of decks (what I call *card game systems*), it had not been ported to my fellow game designer Tim Schutz's alphabet deck: Alpha Playing Cards.[7]

I designed a Hearts game for Alpha Playing Cards using a *subtable* or *internal table* of correspondences that compared the internal features of the standard deck of cards to the internal features of Alpha Playing Cards. For example, I compared the four suits of the standard deck (hearts, spades, diamonds, and clubs) to the two pseudosuits of Alpha Playing Cards (consonants and vowels). I named my game Consonants, because only consonant cards score points in it, just as only heart cards score points in Hearts. This was, in fact, a kenning: I was saying that consonant cards were *the heart cards of Consonants*, and heart cards were the *consonant cards of hearts*.

Thus, a blank space in a table of correspondences showed me an opportunity for a creative act, and I used a subtable to analyze the structural analogies between my starting point (the standard deck of cards and the game Hearts) and my target (Alpha Playing Cards and my new game, Consonants).

I call this technique *analogical design*, and it's powerful. For an even more practical example, consider *The Hitchhiker's Guide to the Galaxy* (H2G2). H2G2 started out as a radio show, was adapted into a series of science fiction novels, then a TV show, a computer game, comic books, and so on. Each of these involved a kind of internal table of correspondences as the author, Douglas Adams, and his colleagues decided how best to translate a joke, character, alien, or bit of technology from one medium to another.

One blank space remaining to be filled was a *Hitchhiker's* movie. A movie was finally made after the death of Douglas Adams, and while it might not have been an artistic success, it has grossed more than $51 million as of August 2005. You might say that the creators of the movie found an empty niche in the memetic ecology and filled it successfully.

End Notes

1. Fauconnier, Gilles, and Mark Turner. 2002. *The Way We Think: Conceptual Blending and the Mind's Hidden Complexities*. Basic Books. An excellent book on analogical thought.

2. Hofstadter, Douglas. 1997. *Le Ton Beau de Marot: In Praise of the Music of Language*. Basic Books. Another excellent book on analogical

thought, this time from a more artistic perspective. Prepare for an "Aha" on every page.

3. Fauconnier, Gilles, and Mark Turner. 2002.

4. Moravec, Hans. 1988. *Mind Children: The Future of Robot and Human Intelligence*. Harvard University Press.

5. Williamson, Craig. 1982. *A Feast of Creatures: Anglo-Saxon Riddle-Songs*. University of Pennsylvania Press. Partially available online at *http://www2.kenyon.edu/AngloSaxonRiddles/Feast.htm*.

6. Bacon, Francis. 1561–1626. *Of Studies*. Quoted in *http://en.wikiquote.org/wiki/Books*.

7. Alpha Playing Cards; *http://www.tjgames.com*.

See Also

• If you enjoy playing with kennings and tables of correspondences, you might enjoy visiting the home page for my game, Kennexions, on the Glass Bead Game wiki, at *http://www.ludism.org/gbgwiki/Kennexions*.

HACK #26 Enjoy Good, Clean Memetic Sex

You can think of conversation as a kind of mental sex that produces ideas rather than physical offspring. To produce good ideas, though, it's best to follow a few rules of "memetic hygiene."

Memes are self-reproducing ideas. According to the theory of memetics, they act like genes by using our minds to replicate themselves, just as our genes use our bodies to do so. The idea of memes was independently discovered by British ethologist Richard Dawkins and several other thinkers. Dawkins, who coined the term *meme*, explains memes this way:

> Examples of memes are tunes, ideas, catch-phrases, clothes fashions, ways of making pots or of building arches. Just as genes propagate themselves in the gene pool by leaping from body to body via sperm or eggs, so memes propagate themselves in the meme pool by leaping from brain to brain via a process which, in the broad sense, can be called imitation.[1]

There are many similarities between genes and memes. Just as genes are transmitted during sexual intercourse in the biosphere, so are ideas transmitted during social intercourse in the mental realm, or *ideosphere*. Table 3-5 shows some examples of correspondences between the genetic and memetic realms.

See "Think Analogically" [Hack #25] for more information on using tables of correspondences.

Table 3-5. Genetic/memetic correspondences

Genetic	Memetic
Gene	Meme
Sexual intercourse	Social intercourse
Genetic engineering	Memetic engineering (for example, marketing and the art of rhetoric)
Seduction	Persuasion
Conception of an embryo	Conception of a new, "embryonic" idea
Sperm bank	Library
Virginity	Ignorance

The memetic realm also has some important differences from the genetic realm. Memes combine, recombine, mutate, and reproduce much more flexibly and rapidly than genes do. This is one way that genetic sex does not map completely to memetic sex. For example, the memetic counterparts of gender and sexual orientation are complicated. From a memetic standpoint, we are all intersexual beings: everyone is able to both transmit and receive ideas, although some people have a stronger tendency toward one than they do toward the other. I'll just suggest that a memetic equivalent of the Kinsey Scale might be called for, and then move on.

In Action

Good, clean, fulfilling memetic sex is the birthright of anyone with a brain, so it's important to educate yourself about memetic sexual hygiene. Here are nine points to keep in mind.

Seek partners outside your family. *Inbreeding* is sex between two individuals that are genetically too similar, and it often produces low-quality offspring. The memetic equivalent of inbreeding (memetic incest) is *groupthink*, which sometimes happens when two or more people who are too similar try to create something.

It's a commonplace idea that Hollywood is *incestuous*: the same people with the same limited set of ideas meet over and over. Is it any surprise, then, that so many Hollywood movies are so bad and that so many of them look so similar? The logical culmination of memetic inbreeding is Jonestown.

Seek partners within your species. Conversely, sometimes people try to memetically mate *outside their memetic species*. From the perspective of one specialist (such as a Chaucer expert), another specialist (such as a particle physicist) is a member of another memetic species.

Because memetics is fluid, it's not true that you can *never* create ideas with someone dissimilar from yourself, because all humans have *some* memes in common. Nevertheless, the rabid anime fan and the rabid model railroad enthusiast are *relatively* mutually infertile.

Broaden your taste. Members of two different memetic species can sometimes spark interesting ideas from each other, and it's good to remain open to that possibility. From the perspective of a generalist or comprehensivist, however, a specialty is less like a species and more like a memetic fetish.

Generalists are flexible; because they don't have fetishes (specialties) themselves, they can enjoy memetic sex with other generalists and also with specialists of all kinds. Generalists are, in short, memetic sluts, and they have a great time, but you don't have to be a memetic slut to benefit from broadening your tastes.

Make sure your partners are healthy. On the other hand, even generalists probably shouldn't have memetic sex with just anyone. You might do well to avoid people in dark suits who ring your doorbell early in the morning with a rabid look in their eyes—and as for the pamphlets people leave at bus stops, well, you just don't know where they've been, memetically speaking.

Practice safe sex. Much as in the biological realm, you can prevent not only unwanted conceptions but also thought viruses in two basic ways: abstinence and condoms.

Abstinence means something like moving to a monastery, taking a vow of silence, and reading nothing but The Book. Wearing a mental condom is more practical and consists of exercising your skepticism, cynicism, irony, and humor.

Gain experience. The memetic virgin, or autodidact, is more likely to produce horrendous doggerel or massive treatises proving that π equals 3 than someone who has had regular memetic intercourse—in this case, an education.

Insist on satisfying sex. The memetic equivalent of orgasm is the "Aha!" moment, which occurs when you completely grok a concept or get a joke. If your memetic partners don't make you say "Aha!" or "Ha ha!" often enough, you might need some new friends, books, TV and radio shows, or other ways to get ideas.

Respect people's boundaries. A *safeword* is a word used during sex that means, "Stop, right now! I'm not kidding!" In real life, the expression "Too much information!" or "TMI!" often functions as a conversational memetic safeword.

Unfortunately, some people have memes that they feel compelled to evangelize at all costs, and they won't stop when they're told to. Memetically, this is the equivalent of rape. Avoid memetic rapists, and respect the boundaries that others set, if you want them to respect yours.

Have lots of kids. As the world moves from using forests for idea storage to ever more efficient electronic devices, there will cease to be any correspondence between the biosphere's population problem and the ideosphere. Thus, although it's not always a good idea in the physical world, in the ideosphere, the more memes (or *mind children*), the merrier!

How It Works

This hack works because genes and memes are *replicators*: forms of information that reproduce and evolve.

Evolution occurs when:

- Patterns reproduce with arbitrary variations.
- The new, varied patterns persist or are destroyed because of environmental conditions.
- The persisting patterns repeat the cycle by reproducing with arbitrary variations.

Since memes share these properties with genes, memes behave in much the same way as genes.

In Real Life

Although author Scott Thorpe does not seem to have encountered the idea of memetics, he has numerous suggestions for fruitful "cerebral sex" in his book *How to Think Like Einstein*.[2] Here are a few suggestions, inspired by his book, from the most risky to the least risky.

Name tags. Write a question you need answered, such as "How can I become a cult author like H.P. Lovecraft?" in big letters on a name tag and wear it everywhere you go. Strangers will go out of their way to answer your question when they see you on the street. Maintain your sense of humor. (See "Practice safe sex" earlier in this hack.)

Memetic orgies. Invite the smartest people you know to a problem-solving party. Inform them that the best beer and munchies will be forthcoming only *after* they've collectively generated some possible solutions for your problem.

Old friends. Old friends are like old lovers, memetically speaking. You probably share a lot of memetic offspring with your old friends, but time has changed you all, and now they have some new memes to share with you—memes that you might very well have conceived yourself had you been in their place.

Memetic marriage. The safest memetic sex of all—and yet often the most satisfying—can be that with a long-term memetic partner, someone who knows you well enough to tickle your memetic toes but is still different enough from you to complement your weaknesses so that your mind children will have hybrid vigor.

Your memetic spouse can be your real-life spouse (mine is), but need not be. For example, two of the best game design partners I know are brothers, and good friends can work well, too.

End Notes

1. Dawkins, Richard. 1989. *The Selfish Gene*. Oxford University Press.
2. Thorpe, Scott. 2002. *How to Think Like Einstein*. Barnes & Noble Books.

HACK #27 Play Mind Music

You can condition yourself to use your favorite music as a creative trigger, as well as a filter for environmental noise, and if you put it on your iPod or MP3 CD player, you'll always have it at hand.

Practically everyone has worked to music at one time or another, from late-night college study sessions to the night shift at a fast-food joint. Consider the beat of a drum to encourage a crew pulling at oars, or the fife and drum corps of an army. Music has always been used to change people's moods and get them to work together; it coevolved with ritual drama, which was also used to change people's state of mind.

However, what is appropriate for one situation might not be appropriate for another, and the same music that gets a team working in synchrony might be completely distracting for someone who is trying to think in solitude. The converse holds true, of course. Some kinds of music are appropriate for thinking that would probably be a drag on physical work.

In Action

Get yourself a good pair of headphones that are as sonically isolated as possible, and make yourself a *mind music* MP3 CD or playlist on your iPod that you can fill with music to help you concentrate and focus on thinking. You might already have an idea what music this might be; if not, see the "In Real Life" section of this hack for some ideas.

Condition yourself to think while this music is playing so that it becomes a kind of musical thinking cap: put it on and you become a thinker for the duration. To that end, listen to this music only when you intend to think so that your conditioned response of thinking hard while it is on does not become extinguished accidentally. You don't need a lot of different pieces in your mind-music collection, because your response will probably be stronger if their scope is limited.

 Think of your unconscious mind as being like your dog. Imagine if you had to teach your dog 100 different words for *sit*.

You must decide how much music should be in your mind-music collection. A short, quickly repeating loop of music drives some people crazy. However, if you have a good sense of how long one of your typical working sessions is, you can use your mind music as a timer. You can pace your work like a workout, with both warm-up and cool-down music. When you hear a certain piece of music, you will know that it is time to start wrapping up.

A short music loop can nevertheless be useful, because you will sometimes associate certain thoughts you are having with a certain part of the music you are hearing, so that on the next time through the loop, you will be reminded of those thoughts when you hear that part of the music again, in an interesting—and sometimes creatively fruitful—mental echo effect.

Instrumental music is preferable, because research has shown that people need to keep the speech centers of their brains free to think about complex information. As pointed out in "Talk to Yourself" [Hack #62],[1] you can listen to pop music or talk radio while driving a familiar route without many turns, but if you get off the highway and need to locate a new destination, you'll turn off the radio because it's distracting you from thinking.

But don't rule out music with lyrics entirely. If you're in a good mood and the music is upbeat, you might find the music will occasionally toss up lyrics that seed your creative random-number generator [Hack #19] as well as get your toes tapping, so you are filled with new ideas.

Chants (such as Gregorian chants) and music in foreign languages are a gray area in this respect. To some extent, you are likely to hear a voice in another language as just another instrument, or your brain might treat the music as a verbal Rorschach test onto which to project some creative and interesting lyrics of its own. Yet another alternative is that you will unconsciously strain to make out the foreign-language lyrics in your own language so that they interfere almost as much as lyrics you do understand. Gauge your own reaction, which will probably vary from piece to piece.

In Real Life

Here are the pieces on my mind-music CD, all instrumental:

- *The Goldberg Variations* by Johann Sebastian Bach, as performed by Glenn Gould
- *Music for Airports* by Brian Eno
- *Neroli: Thinking Music, Part IV* by Brian Eno
- *Suite for Flute and Jazz Piano* by Claude Bolling and Jean-Pierre Rampal

Specifically, I have found that almost any Bach is good for thinking, because of its formal nature. I chose Glenn Gould's performance of the Goldberg Variations because it is legendary; the confluence of a composer and a performer of genius spurs me on. I placed the two fugues on the CD into a separate folder so that I can put just the Goldbergs on repeat mode if I wish; they already form a kind of musical closed circle, so this is appropriate.

Almost at the other end of the spectrum, *Suite for Flute and Jazz Piano* bubbles with playfulness. It is less subdued than the other pieces on my disc and therefore more suited for filtering out noise from the environment. I listen to it when I am designing games, or when I need some playful energy; not only have I loved this piece for a long time, but also I have fond memories of a late friend at Yale who in the spring used to play its flute part barefoot in the courtyard. He was one of the most creative people I ever knew, and I like to think of his playful spirit occasionally entering my work. Thus, one can choose mind music for personal reasons as well.

If you are not familiar with ambient music or Brian Eno, become so. Ambient music is not to all tastes, but as the title *Thinking Music, Part IV* suggests, it is specifically designed to enhance rather than detract from one's ability to concentrate.

As Brian Eno writes in his 1978 "Ambient Music Manifesto:"

> Whereas conventional background music [e.g., Muzak] is produced by stripping away all sense of doubt and uncertainty (and thus all genuine interest) from the music, Ambient Music retains these qualities. And whereas their

intention is to 'brighten' the environment by adding stimulus to it (thus supposedly alleviating the tedium of routine tasks and levelling out the natural ups and downs of the body rhythms) Ambient Music is intended to induce calm and a space to think.[2]

As for the two Eno pieces in my collection, *Thinking Music, Part IV* is deep and somewhat dark, suitable for ponderous thoughts, and *Music for Airports* is somewhat lighter. I note with satisfaction that many other people online have chosen this piece as "thinking music," too.[3,4]

End Notes

1. Stafford, Tom, and Matt Webb. 2005. *Mind Hacks*. O'Reilly. ("Talk to Yourself" [Hack #62] originally appeared in *Mind Hacks* as Hack #61.)

2. Eno, Brian. 1978. "The Ambient Music Manifesto." *http://www.elemental.org/ele_ment/said&did/eno_ambient.html*.

3. Mentat Wiki "Mentat Music" page; *http://www.ludism.org/mentat/MentatMusic*.

4. "Good Thinking Music" pages on the original wiki; *http://c2.com/cgi/wiki?GoodThinkingMusic* and *http://c2.com/cgi/wiki?GoodThinkingMusic Testimonials*.

HACK #28 Sound Your Brain with Onar

Onar, or oneiric sonar, is a hack for plumbing your unconscious mind in search of new ideas during hypnagogic sleep. It's similar to methods used by Salvador Dali and Thomas Edison.

Hypnagogia is the mental state between waking and sleep, the "half-asleep" state. The word comes from the Greek *hypnos* (sleep) and *agogeus* (leader or conductor); hypnagogia is the state that leads us into or out of sleep.

> Some researchers draw a distinction between *hypnagogia*, which occurs when we are falling asleep, and *hypnopompia*, which occurs when we are waking up.

Many thinkers throughout history have found hypnagogia to be a fathomless well of creative black gold. For example, the surrealist painter Salvador Dali developed a technique to help him visualize dream landscapes of bizarre beauty, which he would paint upon awaking.

Dali is also said to have trained himself to doze in a chair with his chin resting on a spoon that was held in one hand, propped by his elbow, which rested on a table. In this position, when his muscles relaxed and he was on the verge of falling asleep, his chin would drop and he would wake, often in the middle of a hypnagogic dream or vision which he would then proceed to paint.[1] For instance, this technique likely inspired his painting "Dream Caused by the Flight of a Bumblebee Around a Pomegranate a Second Before Awakening."[2]

Such techniques can also be useful to hardheaded businesspeople and inventors. For example, Thomas Alva Edison was known to use a similar technique. He put the hypnagogic state to work when he was an adult and had an unusual technique: he would doze off in a chair with his arms and hands draped over the armrests. In each hand, he held a ball bearing. Below each hand on the floor were two pie plates. When he drifted into the state between waking and sleeping, his hands would naturally relax and the ball bearings would drop on the plate. Awakened by the noise, Edison would immediately make notes on any ideas that had come to him.[3]

A partial list of other thinkers who have been inspired by hypnagogia is impressive:[4]

Artists
Jean Cocteau, Max Ernst

Musicians
Johannes Brahms, Giacomo Puccini, Richard Wagner

Poets
William Blake, John Keats

Scientists
Albert Einstein, Friedrich August Kekulé

Writers
Ray Bradbury, Charles Dickens, Johann Wolfgang von Goethe, Edgar Allan Poe, Robert Louis Stevenson, Leo Tolstoy, Mark Twain

I call the techniques used by Dali, Edison, and some of these other thinkers *onar*, which is a portmanteau [Hack #50] of *oneiric* ("related to dreams") and *sonar*. Onar allows you to sound your mind in sleep the way a submarine sounds the depths of the ocean. This hack shows how to do it without spoons, ball bearings, or pie plates. You can even do it on an airplane.

In Action

To undertake an onar expedition, you will need a quiet place to doze and some way to take notes (in the dark, if necessary). The following seven steps comprise the onar process:

1. Lie on your back in bed or sit in a comfortable armchair.

2. Rest your elbow on the surface of the bed or the arm of the chair so that your forearm is pointing straight up. Let your wrist go limp if that is more comfortable for you.

3. Focus your mind on a problem you wish to solve.

4. Allow yourself to drift toward sleep, while continuing to focus on the problem as long as you can.

5. Wait for your arm to relax and fall, waking you up. This will happen naturally when you begin to fall more deeply asleep.

6. Record any creative thoughts you had while dozing.

7. Repeat.

Depending on how sleepy you are, your arm might not so much fall as merely twitch. Don't worry about it; as long as you can enter hypnagogia and bring back interesting thoughts and images, onar should work for you.

How It Works

Hypnagogia occurs when you're falling asleep. As you fall asleep, your brain turns inward, shifting from the alpha and beta waves of ordinary waking consciousness and sensory processing to the deeper delta and theta waves of later stages of sleep, including dreaming.[5]

With few or no sensory anchors during hypnagogia, the sleeper begins to hallucinate. (I hypothesize that a similar phenomenon occurs during sensory deprivation experiments in "float tanks.") As sensory references disappear, conceptual boundaries also tend to be blurred or lost. This means that ego boundaries are lost, and frames of reference might be blended. Many cognitive psychologists today believe that the blending of conceptual frames is the primary mechanism of creativity.[6] (See "Think Analogically" [Hack #25] for more details.)

In Real Life

I decided I would employ onar on the topic of onar itself, as an example for this hack. I laid on my back one night before sleep with my catch notebook [Hack #13] and flashlight by my side. Actually, I always have them there, because I find that many good ideas come to me during hypnagogia, regardless of whether I want them to.

The first result I obtained while focusing on the concept of onar was the word al*arm*, signifying my current use of my *arm* as a kind of *alarm* clock.

As I fell deeper asleep, I "saw" a diagram like the one in Figure 3-3 that compared using onar to keeping a dream journal [Hack #29].

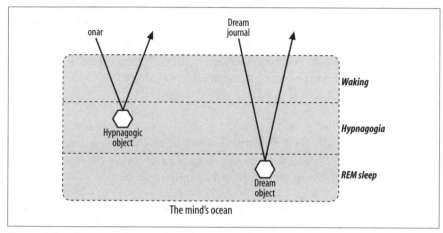

Figure 3-3. A hypnagogic image

The last interesting result I obtained did not relate to this onar hack, but to the Hotel Dominic [Hack #7] instead. It was the idea of using the Hotel Dominic to keep a mental file on each of my friends and acquaintances, containing not only their spouses' and children's names, ages, and so on, but also any personal preferences I might not ordinarily remember, such as whether they like people to remove shoes when entering their houses.

After that short dip into hypnagogia, I relaxed and let myself fall fully asleep, satisfied I had brought back some interesting specimens from the mesopelagic zone of the mind's ocean.

End Notes

1. Mavromatis, Andreas. 1987. *Hypnagogia: The Unique State of Consciousness Between Wakefulness and Sleep*. Routledge & Kegan Paul. An amazing accomplishment, this book is probably the most comprehensive work on hypnagogia in English.

2. URL for the Dali painting: *http://www.artquotes.net/masters/salvador-dali/flight-of-a-bumblebee.htm*.

3. Goleman, D., and P. Kaufman. 1992. "The Art of Creativity." *Psychology Today*. *http://cms.psychologytoday.com/articles/pto-19920301-000031.html*.

4. This list of thinkers was culled from Mavromatis.

5. "Stages of Sleep." *Psychology World. http://web.umr.edu/~psyworld/sleep_stages.htm.*

6. Fauconnier, Gilles, and Mark Turner. 2002. *The Way We Think: Conceptual Blending and the Mind's Hidden Complexities.* Basic Books.

HACK #29 Keep a Dream Journal

Record your dreams to see and hear a rich, creative world.

Dreams can be amazing things and can offer incredible creative riches in the form of beautiful images, plotlines for fiction, even ideas for new inventions. If you want a connection to the activities of the dream world, this is the easiest way to do it.

In Action

Keeping a dream journal is very simple. Just place a piece of paper next to your bed. When you wake up, write down key words about any dreams you had. If you had no dreams, write down "no dreams last night" on the paper—first thing, right out of bed.

That's really all there is to it, and it always works. If you're getting a lot of "no dreams last night," however, when you wake up, don't immediately leap to write "no dreams last night." In fact, *don't move at all.*

Instead, as soon as you realize that you are awake, start searching your mind for dreams. If you can find some leftover fragment of a dream, hold on tight to it. Mentally probe it; you should be able to get some more details. If you're lucky, the whole dream will recur to you at once.

In your mind, start taking notes. (Don't move! Don't get out of bed!) Find keywords to describe what you are experiencing, and memorize them; you don't want to forget. And you *can* forget: if you get up too quickly, you'll forget everything, except the sensation of having looked through a nondescript dream.

Keep exploring your memory of the dream, and keep taking mental notes. When you are confident you have the major parts of it, string the keywords into a sentence in your head. Repeat it over and over and over, just as when you hold a seven-digit phone number in your head, so you can write it down. Mnemonic techniques, such as the number shape system [Hack #2], can be handy here.

Then move, get out of bed, find your pen right next to your journal, and write down those keywords—immediately! Quickly start writing details.

The dream should remain in your mind. Then, write out more details—full sentences, paragraphs, whatever you have time for. If you ride the bus, keep writing what you can remember.

Your memory of the dream should now be fully integrated into your working daily memory.

In Real Life

I like to draw annotated *maps* of places that I've seen in dreams and *pictures* of scenes and peculiar objects I saw. If you can remember the full text of a dialogue, write it out; I once had a good joke told to me in a dream!

Every so often in my life, I reinvestigate my dreams. First, I start remembering a dream a night. After a week or two of paying attention to them, I remember two dreams a night. After another week, I remember several dreams per night. I usually have to stop around here: I start waking up several times in the night, because I keep waking up immediately after (or during) dreaming and have to write them down. It interferes with sleep, and I decide to stop investigating my dreams.

When it gets to be too much for you, and you just want a good night's sleep, just stop paying attention to your dreams. Poof! Uninterrupted nights of sleep.

See Also

- Cartan, John. "Who Do We Dream Against?" *http://www.cartania.com/ dreaming.html.*
- Garfield, Patricia L. 1976. *Creative Dreaming.* Ballantine Books.

—*Lion Kimbro*

H A C K #30 Hold a Question in Mind

One of Sir Isaac Newton's many discoveries was that often to arrive at the truth, you need only contemplate the question.

Besides discovering the principles of gravitation, Sir Isaac Newton discovered a basic principle of human thought: if you want an answer to a question, simply hold the question firmly in mind. When Newton was asked how he had discovered the laws of gravitation, he answered, "By thinking about it day and night." He also said, "If I have ever made any valuable discoveries, it has been due more to patient attention than to any other talent," and "I keep the subject constantly before me and wait 'till the first dawnings open slowly, by little and little, into a full and clear light."[1]

In his book *The Laws of Form*, mathematician and philosopher G. Spencer Brown writes about Newton's insight in this regard:

> To arrive at the simplest truth, as Newton knew and practised, requires *years of contemplation*. Not activity. Not reasoning. Not calculating. Not busy behavior of any kind. Not reading. Not talking. Not making an effort. Not thinking. Simply *bearing in mind* what it is one needs to know. And yet those with the courage to tread this path to real discovery are not only offered practically no guidance on how to do so, they are actively discouraged and have to set about it in secret, pretending meanwhile to be diligently engaged in the frantic diversions and to conform with the deadening personal opinions which are being continually thrust upon them.[2]

The more intently you hold the question in mind, the closer the answer to the question will come to you.

However, Newton's quotations and Brown's commentary require interpretation. People frequently reply, "It can't be like that. Just holding a question in mind doesn't help. You have to *do* stuff."

That is true. You do have to read. You have to work and think. You have to pay attention. Most importantly, you have to be ready to receive results, instead of pursuing them aggressively.

In Action

The key to this hack is that by repeatedly asking the question, you set your mind in such a way that you are receptive. When something relevant crosses your path, your mind will catch it, because it's seeing the world in terms of that particular question. By framing your experiences with the question, you have created a context for the answer to fit into when it comes your way, as in the following examples:

- If you are looking at Google, for instance, the first thing you might think is "Well, can this answer my question?" Ideas of how to search will occur to you, and you can try them.

- If you are at a party, and you have your question in mind, your conversations will gravitate toward the question and its possible answers. You can look for people who may answer your question.

- If you are reading the newspaper, your choice of reading and interpretation of what you read will be shaped by your question.

- If you keep a notebook, keep notes on what you've learned that can help answer the question; putting many pieces of information in one place can help outline a bigger picture.

How It Works

You might not need to put a whole lot of effort into using this hack; your mind will probably provide the suggestions to pull you in the right direction, as long as you are focusing on the question. You'll build curiosity and attentiveness as you stay focused and turn the question around to see all its sides.

"Well, that's obvious," you might say. "I ask questions, I research, I find the answer."

OK, but the point of this hack is twofold: to increase your receptivity and help you think about how many ways a question can be answered, and to help build your confidence, because confidence is another key to finding answers.

When we get frustrated, we are usually in a complex situation. Questions regularly have complicated answers, and it's not unusual that we find ourselves lost in complexities. But complexities can be frustrating.

Confidence and believing that an answer *will* come can help cut through all that. This hack is *simple*. By reminding yourself that "all you have to do" is hold the question in mind, you can relax, perhaps put active investigation (or worry) on hold, and let the mind and the world do whatever it needs to do to help you answer the question. *Don't give up* on the question, but, rather, just look at the world freshly, holding the question in mind. Sometimes just that little bit of relaxation helps a lot.

When you relax, you might notice something from outside of your usual way of thinking that helps you see something new that might lead to your answer.

With time, everything will show itself to you. Somehow, it will all piece together.

Time limits on your seeking, though, are an entirely different matter.

End Notes

1. Ask MetaFilter. 2005. "Who advised people to simply hold important questions in their minds?" *http://ask.metafilter.com/mefi/19491*.

2. Spencer-Brown, G. 1994. *The Laws of Form*. Cognizer Co.

See Also

- When holding a question in mind isn't enough, consult Ask MetaFilter (*http://ask.metafilter.com*). That's what we did to track down the Newton quotations. Note that like another of Newton's discoveries, the calculus, the principle in this hack has been discovered more than once; the earlier MetaFilter thread quotes several other people with similar ideas.

—*Lion Kimbro*

HACK #31 Adopt a Hero

Break down the walls of what you think of as reality; you might find some interesting solutions. A problem to you might not be a problem to someone you adopt as a hero.

Suppose you are writing a screenplay. Here are some examples of questions you might ask yourself when you are stuck:

- With my character Fred's luck, what would happen next?
- What would happen next in a 1940s movie musical?
- What would happen next in a dream?

Some of your most fruitful thinking can occur when you deliberately switch to another way of looking at things. You can think of these switches as being like key changes in a piece of music. For example, you can obtain useful effects by unexpectedly switching "keys" in the middle of a work of fiction from space opera to soap opera.

Modulating from the sublime (such as sainthood) to the ridiculous (such as platform shoes in a 1970s disco) is a comic effect known as *bathos*, but it's equally possible to modulate from the ridiculous back to the sublime. James Joyce's book *Finnegans Wake* often fuses chords of the sublime, the ridiculous, and the grittily political not just within a paragraph or a sentence, but often within a single word. (You can, too, if you put your words in the blender [Hack #50].)

Changing conceptual keys works for all forms of art—at least, in some situations. Imagine that you're an architect or an interior decorator. You might decide that you want the entry to a house to have Gothic sweep, but that a more intimate interior meditation room should be styled after a Japanese *zendo*.

Be conscious of the appropriateness of your borrowings; mixing too many styles or styles that are too discordant can result in a postmodernist mush.

The most useful key change for problem solving, as well as certain art forms, might not be adopting another style, but adopting the entire worldview of another person, creature, or even inanimate object.

In Action

Let's change the key of this hack from music to religion. Consider the *short-duration personal savior* (ShorDurPerSav).[1] By temporarily (for a *short duration*, as opposed to the rest of your life) adopting the viewpoint of another person, even an imaginary one—or one you would ordinarily find repugnant—you can learn much.

Why limit yourself to asking, "What would Jesus do?" (WWJD?) when you can not only ask "WWBD?" ("What would Buddha do?") or "WWMD?" ("What would Mohammed do?"), but also:

WWBBD?
   What would Bugs Bunny do?

WWMAD?
   What would Marcus Aurelius do?

WWMPD?
   What would Mary Poppins do?

WWRMSD?
   What would Richard M. Stallman do?

WWSOHD?
   What would Scarlett O'Hara do?

WWYMD?
   What would your mom do?

Get all New-Agey for a minute and channel that person. Although this hack is related to the old occult idea of the *magical personality*, you don't have to believe that you're literally in contact with another being magically; what you're after is the state of mind you're in when you see a movie, or when you act in one. What would that person do? How would she solve your problem?

When you watch *Gone with the Wind*, you don't literally believe that Scarlett O'Hara exists, but a chill goes through you (admit it!) when she raises her fist to the sky and cries, "As God is my witness, I'll never be hungry again!" Most of the time, you're empathizing with her, feeling what she's feeling, but you can go a step further: you can emulate Scarlett in the same way that a desktop PC circa 2005 can emulate a Commodore 64 circa 1980. The worldview of another person, even (or especially) a fictional one, is like a program that your brain can run.

Try it now. How would each of these people or characters react to finding that they had a flat tire?

What would Bugs Bunny do?
> He'd be undaunted, with absolute faith in himself. If he didn't have a spare, he might trick someone into giving him hers or find another way to get there. And if he couldn't do that, he'd sit down in the shade and whip out a banjo to pass the time.

What would Marcus Aurelius do?
> He'd calmly fix the tire or, failing that, endure the flat tire without sinking into crippling self-pity or anger at the uncaring gods.

What would Mary Poppins do?
> She'd make fixing the flat tire into a game!

What would Richard Stallman do?
> He'd appeal to random passersby on ethical grounds to share their individual talents and help him fix the tire.

What would Scarlett O'Hara do?
> The young Scarlett would use her charm to get someone else to fix it. The older Scarlett would steal someone else's tire if she didn't have a spare, using brute force if necessary.

What would your mom do?
> I don't know; what *would* your mom do?

How It Works

Both method actors and psychotherapists have refined the technique of adopting another person's viewpoint.

Actors must learn how to become another person during a performance. Part of the technique of *method acting* is deliberately not interfering with the process, but letting the alternative personality come through and not censoring its thoughts. Preparation for method acting involves accessing actors' personal memories of the emotions that must be brought forth to give the performance life. They draw on their inner resources to call forth emotions they have experienced at other times and places, and then transform them into the words and motions of the characters they're playing.[2]

Psychotherapists sometimes use role-playing to *feel* their clients' emotions instead of intellectualizing about them. This helps the therapists understand their clients better. Having a supervisor and a set procedure for *de-roling* afterward enables a therapist to go deeply into the experience.[3]

In Real Life

The next time you're confronted with a daunting 11-page tax form, try method-acting a tax attorney. Knowing that there are people who don't mind such forms doesn't make the task of filling them out easy, but adopting the mindset of an attorney who sees such forms every day, and who is anything but intimidated by them, can help.

Such confidence might even help you find more advantageous ways to fill out the forms that you might not have thought about. You might also want to adopt another personality trait: eagerness to finishing the task. If you know someone named Fred with that personality trait, go ahead and ask yourself, WWFD?

One of my friends role-played a bird and learned that wing beats are synchronized with a bird's breathing, which turned out to be true. Albert Einstein contributed to his theories of relativity by imagining he was a photon, and Jonas Salk made progress on his polio vaccine by fantasizing that he was a virus.

Whether it's Scarlett O'Hara's spunk or the knowledge of what it's like to travel at the speed of light, by pretending to be what we're not and switching to other modes of thought, we can often gain access to information and abilities that we don't normally use.

End Notes

1. Dobbs, J.R. "Bob". 1987. *The Book of the SubGenius*. Simon & Schuster.
2. Stanislavski, Constantin. 1986. *An Actor Prepares*. Methuen.
3. Brearly, G., and P. Birchley. 1986. *Introducing Counselling Skills and Techniques: With Particular Application for the Paramedical Professions*. Faber & Faber.

See Also

- Wilson, Robert Anton. 1980. *The Illuminati Papers*. And/Or Press. Each article in this book is written from the viewpoint of a different character from Wilson's fiction.
- Edward de Bono's book *Six Thinking Hats* is another example of "trying on" alternate personalities to solve problems. In this book, the personalities are visualized as hats of six different colors.

Go Backward to Be More Inventive Going Forward

HACK #32

Review the steps that led to a new idea, and maybe you can do even better next time.

When you next have an idea that solves a problem or that you are pleased with for some other reason, take the time to review the process by which you got there. Doing so can help you refine the idea, and it also helps you to have more good ideas in the future.

In Action

One approach is to look at the emotions around idea formation. Try this out the next time you have one of those "Aha!" moments, while the emotions around the discovery are still new. Notice the emotions around the discovery. You might have both positive ones and negative ones. Both are part of your normal discovery process. Use the positive emotions to reward your mind for the connections it has just made. It might feel strange, but do it anyway. This in itself might lead to clarifying related ideas that were just below the surface.

The negative feelings might have been expressed in thoughts like, "I should have seen that sooner." Instead of doing nothing with that feeling, treat it as something useful: a spur to return to what you were thinking just before the idea struck and, if possible, how you felt just before you made the connection. It's important to do this right when you have the idea, while the context of the idea is fresh in your mind and easier to go back to.

If this works for you, the pre-idea thoughts might not be entirely logical; in fact, they can be images or feelings, a sense of what's important, or a fleeting impression that felt *right*. You're not trying to explain the thought process; just reaching it again is enough. Sometimes the idea came because you were also thinking about something else at the same time. Going back to the moments when you made the connection makes you more aware of how you made it, which helps you make similar connections in the future more easily.

How It Works

The rapid change of emotions around a discovery can make it harder to reach the thoughts that happened earlier. Those thoughts happened in a different emotional context. The approach I've described to get back to those thoughts works consciously with the emotions. It uses the energy of the emotions of discovery to get back to the earlier emotional context.

Thought formation is often nonverbal, yet reporting on it tends to lead to a more verbal way of thinking. To counteract this, the hack deliberately looks for the nonverbal connections.

The key aspect of the hack is the conscious act of looking at the process of a discovery at the time of making it, which can be by this route or by a route of your own invention. By giving attention to a process that might normally be taken for granted, you make opportunities to encourage and enhance the process.

The point of reviewing the process of reaching an idea isn't just to understand the process that got you there. It's also to look for an alternative route that could have gotten you there more rapidly or brought you more related ideas. Robert Floyd, winner of the 1978 ACM Turing Award for Excellence, has this to say about his own process of discovery:

> In my own experience of designing difficult algorithms, I find a certain technique most helpful in expanding my own capabilities. After solving a challenging problem, I solve it again from scratch, retracing only the insight of the earlier solution. I repeat this until the solution is as clear and direct as I can hope for. Then I look for a general rule for attacking similar problems that would have led me to approach the given problem in the most efficient way the first time. Often, such a rule is of permanent value.[1]

Floyd's suggestions help streamline the process of discovery. More than that, they also help to clarify how seemingly unrelated discoveries may be related at a deeper level.

In Real Life

I had a small epiphany that led me to a method for estimating the number of wing beats per second for a wasp. The "Aha!" moment was the realization that the pitch of the note I could hear in a wasp's buzz gives an indication of the rate at which it beats its wings. Lower notes are slower; higher notes are faster.

Tracing it back, I found that I didn't arrive at the idea by any logical process or argument, but more intuitively. Part of the idea was remembering that one of the pioneers of computers, Herman Hollerith, taught his clerks to count decks of punched cards by flicking through them and listening to the sound. I was imagining this being done slowly (click-click-click) and then speeded up. Part of the idea was imagining the wasp flying in slow motion, slow enough that I could see the individual cycles of wing movement. The idea of rapid counting brought these two ideas together.

Could I have gotten the idea more quickly? Probably so. If I'd thought about the changes in sound an aircraft engine makes as it slows down, I might also have thought of the relation between the pitch of a sound and speed.

End Notes

1. Bentley, Jon. 1988. *More Programming Pearls*. Addison Wesley. Bentley lucidly explains the principles of some algorithms for generating permutations that Floyd designed and reports this quotation, from Floyd's Turing lecture on "The Paradigms of Programming."

—James Crook

HACK #33   Spend More Time Thinking

Become more productive by staring into space—purposefully.

I've suggested elsewhere that sometimes it's useful not to overthink things [Hack #48]. Sometimes, however, it *is* useful to chew a topic over and over until you know it like the taste of your own mouth.

As part of a recent job, I had to spend up to an hour twice a day commuting on a bus. I spent a lot of time reading. I also spent a lot of time meditating [Hack #60], because if you can meditate on a noisy, crowded bus and not just in a tranquil *zendo*, you're getting somewhere, and I don't mean downtown. But the thing I did the most by far was stare into space, and I don't regret it at all.

What I was really doing was thinking, pure thinking. Thinking is a wonderful way to spend your free moments. Thinking is portable, inexpensive, and environmentally safe; it requires no equipment, and you can do it even with a serious physical disability. Best of all, since thinking is universally applicable, you can make progress on any problem of interest simply by directing your attention toward it.

The kind of thinking I'm describing is extremely focused. I'm not talking about daydreaming here, although experiments at Yale have shown that daydreaming can increase your self-control and creativity.[1]

Recent research suggests daydreaming might have harmful effects as well.[2]

What I am talking about is setting some time aside for an extended course of *directed thought*, which can be hard work but is usually more fruitful than woolgathering.

If you're lucky enough to be self-employed or independently wealthy, you can make your own schedule for directed thought. The rest of us can still grab time on the bus, at the doctor's office, in the car while driving the kids to soccer practice, or in line at the grocery store. If you give yourself the gift of this time for purposeful, directed thought instead of (say) ogling soap opera stars in the tabloids at the checkout counter, you will enrich your life considerably.

In Action

There's not much to the technique of directed thought, but that doesn't mean it's as easy as getting hit by a bus, either:

1. Decide what topic or problem you will think about.

2. Turn the problem over in your mind as if it were a three-dimensional object. Scrutinize it. If you notice a new *angle*, follow it and see where it takes you. Alternatively, if you are thinking about an event or process, move around it in time as well as space: *rehearse* it.

3. Don't let yourself get lost. If you find yourself daydreaming, return your attention to the problem at hand.

Step 3 is the hardest, but if you have some experience with meditation [Hack #60], you can return your attention to your problem as if you were returning your attention to your breath during meditation.

> Do not omit step 3. If you do, you will get on the bus thinking about the plot of your novel and leave it half an hour later thinking about what you should have said to your ex-girlfriend five years ago.

In meditation, you allow thoughts to slip through your mind, but in directed thought, you will probably want to hold on to your good ideas. You can do so with a catch [Hack #13], but I promised you this hack required no equipment, and it doesn't. If your environment won't permit you to use a catch notebook (you're in a bumpy moving vehicle, it's too crowded, it's raining, etc.), you can use a mnemonic system to catch your ideas instead. Good candidates for a *mnemonic catch* include the following:

- The number-shape system [Hack #2]
- The number-rhyme system [Hack #1]
- The Major System [Hack #5]
- A memory journey [Hack #3]

When you arrive somewhere that you can write, transcribe your ideas into your catch; later, move them from your catch into more permanent storage, such as a journal or a project file.

In Real Life

You can go far on a project even if you spend only your commuting time on it. As Ovid (43 BC–17 AD) wrote, "Add little to little and there will be a big pile." I call this technique *ratcheting*, and you will be amazed at how well it works if you're new to this idea.

Over the course of a couple of months on the bus, I accumulated a list of about 400 numbered ideas for my GameFrame game project; about 80 percent of the ideas were useful. Thinking on the bus also helped me feel better about my job by making my commute seem less like wasted time, and less like extra time I had to donate to my job instead of spending on my own projects.

Directed thought can be useful for simulating physical objects, too. Nikola Tesla was famous for being able to leave a simulation of a mechanical part "running," in his head for weeks at a time, then coming back to check it later for wear. I don't know anyone with that power of visualization, but my wife, Marty, has it to a lesser degree. She saves on time and materials for her art projects by trying out various assemblages and techniques in her head before finally putting the pieces together and often ends up with more interesting implementations of her original ideas.

End Notes

1. Singer, Jerome L. 1975. *The Inner World of Daydreaming*. Harper & Row.

2. Bell, Vaughan. "Is daydreaming linked to Alzheimer's?" Mind Hacks blog, *http://www.mindhacks.com/blog/2005/08/is_daydreaming_linke.html*.

See Also

- MacLeod, Hugh. "How to Be Creative." This is an excellent essay on how to make the most of those little moments, by an artist whose medium is cartoons drawn on the back of business cards. (The cartoons in the essay are great, too.) Soon to be a book, apparently: *http://www.gapingvoid.com/Moveable_Type/archives/000932.html*.

- Arnold Bennett's Edwardian self-help classic, *How to Live on 24 Hours a Day*, is an excellent motivational book about how to squeeze more out of the 24 hours every human being is allotted each morning: *http://www.gutenberg.org/etext/2274*.

Extend Your Idea Space with Word Spectra

Visualize word clusters to help create and apprehend new concepts.

When translating text from one language to another, you might need words that seem to lie *somewhere between* the available words. When you apprehend new concepts, you might face a similar problem. Visualizing the spectrum of meanings that lie between two words can help you form new concepts and work with them more easily.

In Action

Here are two examples of foreign words that are useful, but that are almost untranslatable. Please note that you don't need to speak German or Portuguese to appreciate the problem!

From German, we have *Gemütlichkeit*. It's a description of a good mood, the warm feeling of being together with good friends, and it usually also involves wine or beer. How should it be translated? Happiness? Companionship? Smugness? It's none of these and all three. Figure 3-4 shows a visual way to represent the untranslatable.

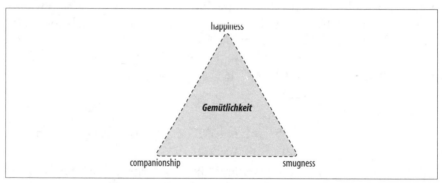

Figure 3-4. Word spectrum for the German word Gemütlichkeit

The second example is once again a word that describes a mood, but this time from Portuguese. The word *saudade* is a mix of homesickness, nostalgia, and good memories, with a tinge of sadness. There simply is no direct translation into English. In one context, it might translate to *nostalgia*, in another to *homesickness*. If you visualize nostalgia and homesickness as being at two ends of a spectrum, *saudade* lies somewhere in between, as shown in Figure 3-5.

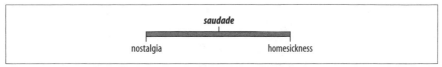

Figure 3-5. Word spectrum for the Portuguese word saudade

So, what about new concepts in English that don't yet have words for them? Biologists run into this problem time and again; evolution just doesn't respect the boundaries made with words.

For example, many of the components in subsystems that repair damage to the body are also part of the normal growth and development system. Repair and growth can be placed on a spectrum. The system that strengthens bones where they are under greatest stress could be classified arbitrarily as being a repair system, because it repairs microscopic damage from mechanical stress.

Alternatively, the bone strengthening could be seen as part of the normal growth process, bringing the right components to the place where they are needed most. It all depends on your point of view. As with the foreign-language examples, it looks as if there is a word missing that combines aspects of growth and aspects of repair (Figure 3-6). The hack of visualizing repair and growth as being at two ends of a spectrum makes it easier to explore the connections between those two concepts.

Figure 3-6. Word spectrum for the English words repair and growth

The value of finding spectra between words isn't confined to biology. In computer programming, recursion and dynamic programming are sometimes taught as if they are two entirely different techniques. In reality, dynamic programming and recursion are at different ends of a spectrum of programming techniques that break larger problems into smaller ones. This is a useful idea to have, because a recursive formulation of an algorithm may be a lot simpler than the corresponding dynamic programming formulation.

How It Works

This hack works because language is inherently a process of using discrete words to cover ranges of meaning. Visualization of word positions, as in the triangle of words for *Gemütlichkeit*, encourages us to see the continuous range of meanings again. We can then home in on meanings that have been missed by the words available to us. The visual position gives us a *handle* for a particular meaning.

The actual visualization of the words doesn't seem to be absolutely essential to the hack. The essential part is identifying that the *endpoints* are related and that there is some possibility between them that combines their elements. However, visualization makes the hack easier to apply, and the visualization helps map words to ranges of meaning.

See Also

- D. Bickerton's *Dynamics of a Creole System* (Cambridge University Press) analyzes so-called *contact languages*, where simpler variants of languages arise. It shows the heroic measures that speakers of such languages must take to express more complex concepts.
- This hack is related to "Put Your Words in the Blender" [Hack #50] and "Learn an Artificial Language" [Hack #51], but it's less complex. You could use it as a precursor to inventing new words.

—*James Crook*

CHAPTER FOUR

Math
Hacks 35–43

Numeracy (the ability to use numbers) is as important as literacy. And while people vary in mathematical aptitude, almost everyone can improve their math skills, because numbers are happy to repay the effort of making their acquaintance.

Because so many people in our culture have trouble with math, improving your math skills will not only help you directly, but can also help you look good and give you an edge at work or school. In keeping with the "mental arts" approach outlined in the Preface, most of the hacks in this chapter involve mental math: math you can do in your head.

Put Down That Calculator

#35

You don't need a calculator to do simple math! Learn a few tricks, and with a little practice, you'll be surprised how much arithmetic you can do in your head

Most people need a calculator to do even simple arithmetic. There's nothing wrong with that, but if a calculator isn't available, it can become a problem. And even if you don't count the time to find the calculator, mental arithmetic can actually be faster than a calculator, too.

In Action

Entire books have been written on mental arithmetic, so we're not going to cover everything in this hack. This hack covers some typical techniques useful in their own right, and some of the other hacks in this chapter are also useful in doing mental mathematics. If you find this hack interesting and useful, you can check out one of the many books on the subject, some of which are listed at the end of this hack.

You should start at a level that's not frustrating for you. If you reach for a calculator to multiply 8 × 7, start by learning the multiplication tables. Use paper and pencil at first, and *check your work*.

Rearrange. Suppose you need to add the following numbers:

    9 + 8 + 7 + 6 + 5 + 4 + 3 + 2 + 1

You *could* add 9 + 8 to get 17, and then add 7 to that to get 24, and so on. But it's much easier to rearrange the addition:

    9 + 1 + 8 + 2 + 7 + 3 + 6 + 4 + 5

Each of the first pairs of numbers adds up to 10. So, we have the following easy addition:

    10 + 10 + 10 + 10 + 5

which is 45.

In addition to rearranging to find 10s (or 20s), rearranging numbers so that they're in descending order tends to help. For instance, suppose you're adding the following numbers:

    1100000
     270000
    3300000
    + 30000

It's probably easier to rearrange that as follows:

    3300000
    1100000
     270000
    + 30000

It's easier because you don't need to keep track of as many nonzero digits while adding 3,300,000 and 1,100,000. Adding 270,000 and 30,000 will also help, so you're left with 4,400,000 + 300,000—an easy sum that totals to 4,700,000.[1]

Put down that burden. When you learned to do paper-and-pencil multiplication in school, you probably learned to work from right to left, carrying as you went:

      ^{12}
       841
    x 74
    3364
    5887
    62234

Notice that you had to carry twice, once in multiplying the 4 in 74 by the 4 in 841, and once in multiplying the 7 by that 4. If you try to do this mentally, you'll have to keep track of multiple numbers between steps. For instance, when multiplying the 7 by the 8, you need to remember the 3,364 from the first multiplication, the 87 you've already figured out, and the carried 2.

Instead of working right to left, let's work left to right, just multiplying one of the six pairs of digits at a time. We're not doing this just to be different; rather, we want to limit the amount of information we need to keep track of.

Before we attempt our previous example, let's do a simpler problem. Multiplying a two-digit number by another two-digit number turns out to be particularly nice. Suppose we need to multiply 42 × 29. We multiply each pair of numbers (keeping track of the powers of 10), starting at the left, and keep a running total:

The individual calculations look like this:

```
40 x 20 = 800
40 x 9 = 360, and 800 + 360 = 1,160
2 x 20 = 40, and 1,160 + 40 = 1,200
2 x 9 = 18, and 1,200 + 18 = 1,218
```

Notice that we have to remember only one number between steps; this remains true even for larger problems. Of course, there's nothing wrong with writing down that number if paper and pencil are handy, and you'll probably find this method is still easier and less error prone that the usual method.

 Because you need the number for only a short time, the mnemonics from Chapter 1 probably aren't necessary here.

Moving from higher numbers to lower ones tends to work better because it's easier to add a small number to a large one. As a bonus, if you need only an estimate, you can stop after doing the first few multiplications.

Now, back to our initial example. It contains pairs of digits:

The calculations look like this:

    800 x 70 = 56,000
    800 x 4 = 3,200, and 56,000 + 3,200 = 59,200
    40 x 70 = 2,800, and 59,200 + 2,800 = 62,000
    40 x 4 = 160, and 62,000 + 160 = 62,160
    1 x 70 = 70, and 62,160 + 70 = 62,230
    1 x 4 = 4, and 62,230 + 4 = 62,234

Of course, this is the same answer that we got before; doing a problem two ways is a good way to check it.[2]

"Calculate Mental Checksums" [Hack #38] and "Estimate Orders of Magnitude" [Hack #41] discuss other ways of checking your work.

Look for friendly numbers. Which addition problem would you rather do: 79 + 87 or 80 + 86? Probably the second; it's easier because the 80, ending in 0, is a friendly number [Hack #36]. For addition, numbers ending in 0 are friendly because adding the corresponding place is trivial (adding 0 to a number doesn't change it). Thus, we can mentally add the tens place (8 + 8 = 16) and then just append the 6 for the ones place to get the answer, 166.[3]

The trick for more difficult addition problems is to change the problem without changing the answer so that we have friendly numbers. For instance, if we had 79 + 87, we'd notice that 79 is near the friendly number 80. To turn 79 into 80, we have to add 1, so to keep the answer the same, we need to subtract 1 somewhere. Let's subtract 1 from the other number, 87, to get 86 and do 80 + 86 = 166, as in the previous example.

Alternatively, you can add 80 + 87 = 167, and subtract the 1 from the result.

The same principle works with multiplication.[4] Suppose I ask you to compute 300 × 70 in your head. Multiplying 3 × 7 = 21 and then adding the three 0s, you easily get 21,000. Again, the 0s at the end make the multiplication easy.

If we need to multiply 302 × 69, we can think as follows:

    302 x 69 = (300 + 2) x (70 - 1)

Now we can do the same cross-multiplication we did before, but with bigger chunks:

```
300 x 70 = 21,000
300 x -1 = -300 and 21,000 - 300 = 20,700
2 x 70 = 140 and 20,700 + 140 = 20,840
2 x -1 = -2 and 20,840 - 2 = 20,838
```

Numbers that end in 0 are the friendliest, but factors of 100 [Hack #36] are pretty friendly, too. For instance, if you think about the factors of 100, you'll find that 25 is a friendly number, and 75 is at least the friend of a friend. Then, the example we used earlier, 841 × 74, looks like this:

```
841 x 74 = 841 x (75 - 1) = 841 x 75 - 841
```

Remember the minus 841, and let's do 841 × 75:

```
841 x 75 = 841 x 3 x 25
         = 2523 x 25
         = (2524 - 1) x 25
         = 2524 x 25 - 25
         = 2524 x 100 / 4 - 25
         = 252400 / 4 - 25
         = 63100 - 25
         = 63075
```

Finally, subtract the leftover 841, part by part:

```
63,075 - 800 = 62,275
62,275 - 40 = 62,235
62,235 - 1 = 62,234
```

which is a third confirmation of the result!

How It Works

All of these tricks rely on basic properties of integers. For instance, the first multiplication example we gave was 841 × 74. We figured that out by using:

```
841 x 74 = (800 + 40 + 1) x (70 + 4)
         = 800 x (70 + 4) + 40 x (70 + 4) + 1 x (70 + 4)
         = 800 x 70 + 800 x 4 + 40 x 70 + 40 x 4 + 1 x 70 + 1 x 4
```

The first line uses the definition of the decimal number system [Hack #40], and the remaining lines are repeated applications of the distributive property.

In Real Life

Suppose you're playing a card game [Hack #67] with a 52-card deck. Your final score is based on your final cards, and each number card is worth its number (so the 4 of hearts is worth 4 points), with an ace worth 1 point and each face card worth 10 points. Your hand is shown in Table 4-1.

Table 4-1. The cards you're dealt

Suit	Cards
Spades	2, 7
Hearts	A, 8
Diamonds	2, 3, 8, 10
Clubs	A, 2, Q

The rearranging method works especially well because we have actual cards to rearrange.

Put the 10 and queen aside: they're worth 20 points. Then, arrange cards in the following groups (Figure 4-1):

- Ace of hearts, 2 of spades, 7 of spades = 10 points
- 8 of hearts, 2 of diamonds = 10 points
- 8 of diamonds, 2 of clubs = 10 points

Figure 4-1. Grouping the cards in your hand

There are 50 points so far; we just need to add the remaining cards, which are the ace of clubs and the 3 of diamonds, for 4 more points. Your total score is 54 points, and you've probably figured this out so much faster than anyone else that you can double-check the rules to make sure the three 2s don't get you some kind of bonus.

End Notes

1. Sticker, Henry. 1955. *How to Calculate Rapidly*. Dover.
2. Julius, Edward H. 1996. *More Rapid Math: Tricks and Tips*. Wiley.

3. Kelly, Gerard W. 1984. *Short-Cut Math*, Chapter 2. Dover.

4. Julius, Edward H. 1992. *Rapid Math Tricks and Tips*. Wiley.

See Also

- Gardner, Martin. 1989. *Mathematical Carnival*. Mathematical Association of America. Chapters 6 and 7 discuss calculating prodigies and some of the tricks they used.

- Doerfler, Ronald W. 1993. *Dead Reckoning: Calculating Without Instruments*. Gulf Publishing Company. This book is a bit more advanced. In addition to the basic operations, it covers extracting roots, and even higher mathematical functions, such as logarithms and trigonometric functions.

—*Mark Purtill*

HACK #36 Make Friends with Numbers

With a little experience, you can learn to recognize many individual numbers by their special properties. Some of these properties can help you with mental arithmetic and memory.

Just like a face in a crowd, a number such as 1,729 probably doesn't mean much to you. But 1,000,000 looks like a friendly face. With some effort, more numbers can look like friends. Here's how to get started making their acquaintance.

In Action

Let's start with some numbers that you're probably already friendly with—10 and its powers: 100, 1,000, and so forth. Because we use a decimal number system **[Hack #40]**, powers of 10 end in zeros. Companies are aware of that and try to get phone numbers that are multiples of a power of 10, because those numbers are easy to remember. For instance, the publisher of this book, O'Reilly Media, has a local phone number of (707) 827-7000, which is a multiple of 1,000. Multiplying by a power of 10 is easy; you just have to append zeros:

```
314 × 1000 = 314000
```

Similarly, dividing by a power of 10 just removes zeros (or moves the decimal point to the left if there aren't enough zeros):

```
2030 / 100 = 20.3
```

If we look at the factors of 10, which are 2 and 5, we can come up with other useful rules. (Factors of 10 are also called *aliquot parts*.) For instance, to multiply a number by 5, first multiply by 10 and then take half the result.

This is based on the notion that, for multiplication and division, 2 is a friendlier number than 5. For instance, $386 \times 5 = (386 \times 10) / 2 = 3,860 / 2 = 1,930$.

This idea can be extended to factors of powers of 10. For instance, $100 = 4 \times 25$, so to divide by 25, double the number twice (which is the same as multiplying it by 4) and divide by 100:

```
217 / 25 = (4 × 217) / 100 = 868 / 100 = 8.68
```

If you're estimating, near factors can also be useful: $33 \times 3 = 99$, which is almost 100, and $17 \times 6 = 102$, which is just a little more than 100.

How It Works

Unlike some of the hacks in this chapter, this one can't give you a straight-forward recipe that's guaranteed to make every integer you encounter unique. After all, the hack involves finding something unusual about the number. If every number you met exhibited the same unusual feature, it obviously wouldn't be unusual. Becoming more familiar with the number system is a gradual process. However, I can give you a few tips:

Start with smallish numbers
Although we'll give an example of hacking a seven-digit phone number later, it might be easier to hack the three-digit and four-digit parts separately, or even split the number into the three-digit part and two two-digit parts.

Try factoring
We already talked about factors of powers of 10 (such as 4 and 25). Factoring in other ways can be useful in mental (or paper-and-pencil) arithmetic. For instance, if you have to multiply 193 by 56, it might be easier to factor 56 into 8×7, and then multiply each factor separately:

```
193 x 56 = 193 x 8 x 7 = 193 x 2³ x 7 = 386 x 2² x 7 = 772 x 2 x 7 =
1544 x 7 = 10808
```

In this example, we multiplied by 8 by doubling three times.

Look for patterns and near patterns
I once had a prescription number that was 66123465. This was easy to remember because of two patterns: all my prescription numbers started with 66 and the rest of the number was almost 123456. Remembering the one difference from the pattern was easier than remembering the whole number.

Associate numbers with uses
If you're a sports fan and you know your favorite player wore number 80, remembering other 80s might be easier by associating them with

that player. (This is a variant of the mnemonic techniques presented in Chapter 1.)

Find friends of friends

In arithmetic, this mostly means numbers close to other numbers. For instance, 4×24 is easy to compute as $4 \times (25 - 1) = 100 - 4 = 96$.

In terms of mnemonics, any way you can connect numbers is fine.

In Real Life

We've already talked about arithmetic, so we'll concentrate on mnemonics here. Let's see what we can do with O'Reilly's toll-free number—(800) 998-9938—in case you want to order more copies of this book. (What am I saying, "in case"?)

Since many toll-free numbers start with 800, this should not be a problem. Notice that 800 was chosen to be a multiple of 100. Then notice that the 998 almost repeats: not only is the second part 99?8, but the extra digit is a 3, which you can think of as a broken 8.

Depending on your tastes, there are other ways to proceed. For instance, back in college, I remember one time I ran into my friend Ben at the beginning of the school year. He told me his phone number: 436-7062. I didn't bother writing it down. I was sure I could remember it without writing it down. And, 20 years later, I do remember it, even though Ben moved out of that particular dorm room 19 years ago.

How did I do it? I saw that number not merely as a sequence of seven numerals, in which form it seems rather arbitrary, but as a particular seven-digit number, and looked for its special properties as a number. Let's try factoring to start, although we'll use some of the ideas mentioned previously as we go along.

First, checking for divisibility [Hack #37], I saw that 4,367,062 is divisible by 2 (because the last digit is even) and 7 (because $+62 - 367 + 4 = -301$ and $-301 / 7 = -43$). I could also see that there were no more small factors. If we divide by the factors I got (that is, 2 and 7) and use techniques similar to those in "Put Down That Calculator" [Hack #35], we obtain the following results:

```
4367062 / 2 = 2183531
2183531 / 7 =  311933
```

So, $4,367,062 = 2 \times 7 \times 311,933$.

Now, there's a pattern to the number 311,933: if you multiply the first three digits by 3, you get the last three digits. A little thought (and a nodding acquaintance with a few other integers) tells us that this means:

```
311933 = 311 x 1003
```

Here I got lucky: both of those numbers were already friends (so our number 311,933 is a friend of those friends). 311 is prime, and 1,003 is 17 × 59, also both prime. As a mathematician, prime numbers—those numbers with no factors other than 1 and themselves—are interesting and often "friendly," and for various reasons[1] 17 is a *very* old friend, so I already knew the factorization of 1,003.

The complete factorization is:

    2 x 7 x 17 x 59 x 311

For me, all of those numbers are friends, and the factorization is nice, with only one of each prime factor. That was enough for me to remember the phone number.

That information might not help you remember the number. The key is to make friends with numbers in your own way. Everyone will have different ways of doing this. For instance, mathematicians love to tell the story of Srinivasa Ramanujan, a great mathematician active in the early 20th century. His friend and colleague, G. H. Hardy, visiting him in the hospital, was making small talk. Hardy mentioned that the cab that took him to the hospital was number 1729—to Hardy's disappointment, because it was such a boring number. Ramanujan disagreed: 1,729 was quite an interesting number, being the smallest number that can be written as a sum of two positive cubes in two different ways.[2]

Ramanujan was quite correct: $10^3 + 9^3 = 1,000 + 729 = 1,729$, and $12^3 + 1^3 = 1,728 + 1 = 1,729$. It's straightforward, if a little tedious, to check that no smaller integer has this property. If I told you the property that Ramanujan mentioned, you could figure out the number 1,729 with a lot of tedious, routine computation, but to go in the other direction is rather remarkable. Even most professional mathematicians would have trouble doing that, except for the fact that this story is so popular. J. E. Littlewood, another colleague and frequent collaborator of Hardy and Ramanujan, put it this way: every positive integer was one of Ramanujan's personal friends. A single face in the crowd is recognizable if you know the person. So it was with numbers for Ramanujan.

Of course, even if you can't recognize *every* face in a crowd, recognizing *some* faces can still be helpful!

End Notes

1. Lefèvre, Vincent. 1998. "17 (Seventeen) and Yellow Pigs." *http://www. vinc17.org/yp17/index.en.html.*

2. MathPages.com. "The Dullness of 1729." *http://www.mathpages.com/home/kmath028.htm*.

—Moses Klein and Mark Purtill

Test for Divisibility

HACK #37

It's often useful to know whether one number is evenly divisible by another number. Here are some tricks that go beyond knowing whether a number is odd or even, or divisible by 10.

Before decimals such as 3.5 were invented, people had to use numbers with fractional parts instead, such as 3½. In many division problems, they had to reduce fractions with large numbers—for example, 243 / 405—to their lowest terms—in this case, ³⁄₅. Knowing rules to determine divisibility by the integers from 1 through 12, or from 1 through 15, was very useful in that precalculator time.[1]

If you want to strengthen your mental math powers, knowing the same rules can be useful to you today. In particular, these rules are helpful with math tricks that involve factoring numbers, such as simplified mental multiplication. Sometimes, knowing *that* a number is evenly divisible by another number goes at least halfway toward knowing *what* the answer is.

In Action

The following list gives tests for divisibility by all integers from 1 to 15. In this context, *divisible* means *evenly divisible*—that is, divisible with a remainder of 0.

1. Every integer is divisible by 1.

2. If the number's last digit is even (0, 2, 4, 6, or 8), the number is divisible by 2. *Examples:* 22, 136, 54, 778.

3. If the number's digit sum is 0, 3, or 6 (or 9, which is the same as 0 for this purpose), the number is divisible by 3. (See "Calculate Mental Checksums" **[Hack #38]** for how to calculate digit sums.) *Example:* 138 (1 + 3 + 8 = 12; 1 + 2 = 3).

4. If the last two digits of the number, taken as a two-digit number, are divisible by 4, so is the number. *Example:* 216 (16 is divisible by 4).

5. If the last digit of a number is 0 or 5, the number is divisible by 5. *Example:* 147,325 (the last digit is 5).

6. If a number is divisible by both 2 and 3, the number is also divisible by 6. (See the tests for 2 and 3.) *Example:* 138 is divisible by 2 because its

last digit is 8. It is also divisible by 3 because $1 + 3 + 8 = 12$ and $1 + 2 = 3$. Therefore, it's also divisible by 6.

7. See the "Divisibility by 7" sidebar.

8. If the last three digits of the number, taken as a three-digit number, are divisible by 8, so is the number. *Example:* 2,889,888 (the last three digits, 888, are divisible by 8).

9. If the number's digit sum is 0 (or 9, which is the same as 0 for this purpose), the number is divisible by 9. (See "Calculate Mental Checksums" [Hack #38] for how to calculate digit sums.) *Example:* 41,805 ($4 + 1 + 8 + 0 + 5 = 18$; $1 + 8 = 9$).

10. If the last digit of a number is 0, the number is divisible by 10. *Example:* 99,310 (the last digit is 0).

11. Casting out elevens [Hack #38] is the easiest way to test for divisibility by 11 in most cases: if the number modulo 11 is 0, the number is divisible by 11.

12. If a number is divisible by both 3 and 4, the number is also divisible by 12. (See the tests for 3 and 4.) *Example:* 624 is divisible by 3 because $6 + 2 + 4 = 12$ and $1 + 2 = 3$. It is also divisible by 4 because the last two digits (24) are divisible by 4. It is therefore divisible by 12.

13. If 9 times the last digit of the number, subtracted from the number with its last digit deleted, is divisible by 13, so is the number.[2] *Example:* 351 is divisible by 13 because $35 - 9 \times 1 = 26$. Since 26 is divisible by 13, so is 351.

14. If a number is divisible by both 2 and 7, the number is also divisible by 14. (See the tests for 2 and 7.) *Example:* 65,282,490 is divisible by 2 because it ends in 0. It is also divisible by 7 because it is 7 less than 65,282,497, which we know is divisible by 7 from the example in the "Divisibility by 7" sidebar. Since it is divisible by both 2 and 7, it is divisible by 14.

15. If a number is divisible by both 3 and 5, the number is also divisible by 15. (See the tests for 3 and 5.) *Example:* 3,285 is divisible by 3 because $3 + 2 + 8 + 5 = 18$ and $1 + 8 = 9$. It is also divisible by 5 because it ends in 5. Therefore, it is divisible by 15.

In Real Life

Here's an example of the kind of situation where knowing tests for divisibility will come in handy in real life.

Divisibility by 7

The following procedure is one of the simplest available for testing divisibility by 7:

1. Take the number you want to test, such as *65,282,497*.
2. Split the number into groups of three digits, starting at the right. If the number is written with commas, the splits will be easy to see. Don't worry if the leftmost group has less than three digits. In this case, the groups are *497, 282,* and *65*.
3. Alternately, add and subtract these groups, treated as three-digit numbers. To continue the example, *+497 − 282 + 65 = 280*.
4. The output of this procedure will have the same divisibility by 7 that the original number does. It's easy to see that *280 / 7 = 40*, so 280 is divisible by 7, and it follows that 65,282,497 is, too.

The same procedure will work for figuring divisibility by 11 and 13: simply test the output of this procedure for divisibility by 11 or 13, respectively, rather than 7.

You're at a dinner for 11 people. The restaurant is closing, and everyone agrees to split the bill evenly to save time, but no one has a pocket calculator or PDA handy.

The bill is $419.15, including gratuity. You round this to $419, and cast out elevens. The result is 1, which means that by subtracting 1 from 419, you'll reach a number evenly divisible by 11, which is 418. Quick mental division shows you that everyone owes $38 (418 / 11 = 38), and that if random people around the table contribute some pocket change to make up the difference of $1.15, you can pack up and get out of the restaurant quickly.

End Notes

1. Gardner, Martin. 1991. "Tests of Divisibility." *The Unexpected Hanging and Other Mathematical Diversions*. The University of Chicago Press. An excellent article on divisibility, and a primary source for this hack. Gives the rules for 1 through 12, several additional tests for divisibility by 7, magic tricks involving divisibility, and more, in the wonderful Gardner style.

2. Wikipedia article. "Divisor." *http://en.wikipedia.org/wiki/Divisor*. Gives the rules for 13–15, defines some terminology, outlines some basic principles, and specifies a somewhat elaborate rule for determining divisibility of any integer, in any base, by any smaller integer.

Calculate Mental Checksums

Computers use checksums to ensure that data was not corrupted in transmission. Now your brain can use a checksum for your mental math, with a few easy techniques.

It's important to have some way to check your mental math that doesn't take as long as solving the problem did originally, and ideally is much shorter. It's easy to check your math for the four basic operations of arithmetic (addition, subtraction, multiplication, and division) by calculating *digit sums* for the numbers involved. A digit sum is a special kind of *checksum* or data integrity check. Checksums are used all over the world of computing, from credit cards to ISBNs on books, to downloads you make with your web browser. Now your brain can use them, too.

Finding the digit sum of a number is easy. Just add all the digits of the number together. If the result is greater than 9, add the digits together again. Continue to do so until you have a one-digit result. If the result is 9, reduce it to 0. The result is the digit sum of the original number.[1]

For example, the digit sum of 381 is 3:

```
3 + 8 + 1 = 12
1 + 2 = 3
```

Similarly, the digit sum of 495 is 0:

```
4 + 9 + 5 = 18
1 + 8 = 9 (same as 0)
```

A number's digit sum is actually that number *modulo 9*—in other words, the remainder when that number is divided by 9. See "Calculate Any Weekday" [Hack #43] for a refresher on modulo arithmetic.

This technique is also known as *casting out nines*. Casting out nines and a similar technique known as *casting out elevens* (discussed in the following section) are all you need to check your arithmetic calculations rapidly and to a high degree of accuracy.

In Action

This section shows how to calculate checksums for the four basic operations: addition, subtraction, multiplication, and division. Only integers are used in the examples, but the techniques will work just as well for real numbers as long as they have the same number of decimal places. For example, if you are multiplying 13.52 by 14.6, think of the latter number as 14.60.

Addition. To check your answer after addition:

1. Find the digit sums for the numbers you are adding.

2. Add all the digit sums together.

3. Find the digit sum of the new number, and the digit sum of the answer number.

4. Compare these two digit sums. If the digit sums match, the answer should be correct.

Here is an example of checking addition successfully:

```
  95   9 + 5 = 14; 1 + 4 = 5
+ 42   4 + 2 = 6
+ 22   2 + 2 = 4 ... 5 + 6 + 4 = 15; 1 + 5 = 6
 159   1 + 5 + 9 = 15; 1 + 5 = 6 : OK
```

Here is another example of checking addition:

```
  49   4 + 9 = 13; 1 + 3 = 4
+ 37   3 + 7 = 10; 1 + 0 = 1 ... 4 + 1 = 5
  76   7 + 6 = 13; 1 + 3 = 4 : WRONG (the answer should be 86)
```

The fact that the digit sum of the numbers being added does not match the digit sum of the result tells you that the result is incorrect.

 Digit sums are even easier to calculate if you treat the 9s as 0s right away. Thus, instead of 4 + 9 = 13 in the example, just compute 4 + 0 = 4 and obtain your digit sum in one step.

Multiplication. To check your answer during multiplication:

1. Find the digit sums for the numbers you are multiplying.

2. Multiply them.

3. Find the digit sum of the new number, and the digit sum of the answer number.

4. Compare these two digit sums. If the digit sums match, the answer should be correct.

Here is an example of a correct multiplication:

```
  33   3 + 3 = 6
x 27   2 + 7 = 9 = 0 ... 6 x 0 = 0
 891   8 + 9 + 1 = 18; 1 + 8 = 9 = 0 : OK
```

Here is another example of checking multiplication:

```
   76   7 + 6 = 13; 1 + 3 = 4
 x 14   1 + 4 = 5 ... 4 x 5 = 20; 2 + 0 = 2
 1164   1 + 1 + 6 + 4 = 12; 1 + 2 = 3 : WRONG (the answer should be 1064)
```

The fact that the digit sum of the product of the digit sums of the numbers being multiplied does not match the digit sum of the result tells you that the result is incorrect.

Subtraction. Subtraction is the inverse of addition. Checking a subtraction problem works the same way as checking an addition problem: simply turn the subtraction problem into an addition problem first, as in the following example:

```
    58         26   2 + 6 = 8
  - 26  →    + 32   3 + 2 = 5 ... 8 + 5 = 13; 1 + 3 = 4
    32         58   5 + 8 = 13;  1 + 3 = 4 : OK
```

Division. Just as subtraction is the inverse of addition, so is division the inverse of multiplication. Thus, checking a division problem works the same way as checking a multiplication problem: first turn the division problem into a multiplication problem, as in the following example:

```
    1081 / 23 = 46

       23   2 + 3 = 5
     x 46   4 + 6 = 10;   1 + 0 = 1 ... 5 x 1 = 5
     1081   1 + 0 + 8 + 1 = 10; 1 + 0 = 1 : WRONG
   (the answer should be 47, not 46)
```

The fact that the digit sum of the product of the digit sums of the numbers being multiplied does not match the digit sum of the result tells you that the division result of 46 is incorrect.

False positives. Sometimes casting out nines will not find an error. For example, sometimes errors in two digits will cancel out, as in the following example:

```
      272   2 + 7 + 2 = 11; 1 + 1 = 2
    + 365   3 + 6 + 5 = 14; 1 + 4 = 5 ... 2 + 5 = 7
      547   5 + 4 + 7 = 16; 1 + 6 = 7 : OK?
```

The correct answer is not 547, but 637 (6 + 3 + 7 = 16; 1 + 6 = 7).

Casting out nines will also not find *errors of place* (when the decimal point is misplaced, or the result is otherwise off by a power of 10). Estimating the order of magnitude **[Hack #41]** (roughly, finding the number of digits to the left of the decimal point) will help catch some errors of place, but casting out elevens is even better.

Casting out elevens. Casting out elevens (that is, calculating numbers *modulo 11*) is slightly more accurate than casting out nines. It will also catch errors that casting out nines will not, including errors of place, so it is useful as a cross-check.

To cast out elevens from an integer, simply add all of the digits in the odd places of the number (for example, the ones and hundreds digits, which are in places 1 and 3, counting from the right), then subtract all of the digits in the even places (for example, the tens and thousands digits, in places 2 and 4).

For example, casting out elevens from 5,924 gives the result $4 + 9 - 2 - 5 = 6$.

If your result is greater than 11, cast out elevens from *that* number, and continue doing so until you have a number that's less than 11. If the result is less than 0, add 11 to it until you have a number that is at least 0 and less than 11. To cast out elevens from a sum, total the result of casting out elevens for each number you're adding, and then cast out elevens from that.

The following example is a cross-check of the sum from the previous section:

```
  272   2 + 2 - 7 = -3; -3 + 11 = 8
+ 365   5 + 3 - 6 = 2 ... 8 + 2 = 10
  547   7 + 5 - 4 = 8 : WRONG (the answer should be 637)
```

How It Works

A rigorous mathematical proof that casting out nines by summing the digits of a number will give you that number's remainder when divided by 9 is beyond the scope of this book, but it's fairly intuitive.

First, consider that 0 mod 9 is 0, 1 mod 9 is 1, 2 mod 9 is 2, 3 mod 9 is 3, and so on. 9 mod 9 is 0, 10 mod 9 is 1, 11 mod 9 is 2, and the cycle continues.

Next, consider that 20 mod 9 is 2, 30 mod 9 is 3, and so on; check and see. You will also find that 200 mod 9 is 2, 2,000 mod 9 is 2, 20,000 mod 9 is 2, and so on. In fact, any integer multiplied by any power of 10 and then calculated modulo 9 has the same result as the original integer modulo 9.

Since $(a + b + c)$ mod 9 is the same as $(a$ mod $9) + (b$ mod $9) + (c$ mod $9)$, and since (for example) the number 523 can be written as $500 + 20 + 3$, simply adding the individual digits of 523 $(5 + 2 + 3)$ will serve the same purpose as calculating the sum of 500 mod 9, 20 mod 9, and 3 mod 9, to wit, finding 523 modulo 9.[2]

Figuring out why casting out elevens works is left as an exercise for the mathematically minded mind-performance hacker.

In Real Life

Mental checksums are most useful when you combine them with other math hacks [Hack #75]. Since checksums using both 9 and 11 are easy to find, checking your work won't add much time to your mental calculations unless you made a mistake—in which case, better late than wrong.

End Notes

1. Julius, Edward H. 1992. *Rapid Math Tricks and Tips*. John Wiley & Sons, Inc.

2. Menninger, Karl, and E. J. F. Primrose (trans). 1961. *Calculator's Cunning: The Art of Quick Reckoning*. Basic Books, Inc., Publishers.

HACK #39   Turn Your Hands into an Abacus

You might have heard stories of how rapid and accurate calculations with an abacus can be, but did you know that the abacus might have been based on an ancient technique using only the human hand, which survives today as the Korean art of Chisenbop?

Chisenbop is an ancient Korean technique for calculations with the human hand. The classic text on Chisenbop in English is *The Complete Book of Fingermath*.[1] Unfortunately, it is expensive, aimed toward children, and takes hundreds of pages to explain principles that an educated adult can learn in a few minutes. One important thing that the book can offer you, however, is page after page of drills. Chisenbop should become a motor skill, not something you have to think about.

Fingermath also uses many full, detailed drawings of hands in action, which is another reason the book is so long. Fortunately, the Wikipedia presents a notation that can radically compress Chisenbop diagrams on the page, as shown in Table 4-2.[2]

I added a couple of symbols to the notation myself for this book (^ and v, which are described in Table 4-2).

Table 4-2 . Chisenbop notation

Notation	Meaning
-	A thumb in the air
@	A thumb with its tip pressed to the table
.	A finger in the air
o	A finger with its tip pressed to the table
^	Lift that finger
v	Press that finger down

You can combine the finger notation across two hands, as shown in the examples in Table 4-3.

Table 4-3. Examples of Chisenbop notation

Notation	Meaning
....‾ ‾....	Both hands free
oooo@ @oooo	All thumbs and fingers down
....@ @....	Thumbs down only
oooo- -oooo	Fingers down only
....‾ -o...	Right index finger down only
...v- ^^^^^	Press your left index finger, and lift all the fingers and the thumb on your right hand

This remainder of this hack describes the basic operations of Chisenbop.

Counting

Here is how to count to 100 or more on your fingers. Keeping all your other fingers off the table, you press your right index finger to the table. This represents the number 1:

Notation	Meaning
-o...	1

Keeping your right index finger on the table, you press your right middle finger on the table. This represents the number 2:

Notation	Meaning
-oo..	2

The next two fingers down represent 3 and 4, respectively:

Notation	Meaning
-ooo.	3
-oooo	4

To represent 5, *clear* your other four fingers on your right hand (lift them off the table) and simultaneously press your right thumb to the table. Your thumb represents 5:

Notation	Meaning
v^^^^	Clear fingers, press thumb
@....	5

For 6, keeping your right thumb on the table, press your right index finger down on the table (5 + 1 = 6):

Notation	Meaning
@o...	6

You can get as high as 9 on your right hand by pressing the next three fingers on your right hand in succession so that all the fingers are pressed:

Notation	Meaning
@oo..	7
@ooo.	8
@oooo	9

To count from 10 up through 99, you will need to start using your left hand as well.

Clear all the fingers on your right hand and simultaneously press the index finger of your left hand. This is 10. It's analogous to how you count 1 on your right hand by pressing your right index finger.

Notation	Meaning
...v- ^^^^^	Clear right hand, press left index finger
...o- -....	10

By adding the fingers of your right hand from 1 through 9, you can get as high as 19:

Notation	Meaning
...o- -o...	11
...o- -oo..	12
...o- -ooo.	13
...o- -oooo	14
...o- @....	15
...o- @o...	16
...o- @oo..	17
...o- @ooo.	18
...o- @oooo	19

Of course, to count to 20, you make another exchange. Clear your right hand and press the middle finger of your left hand:

Notation	Meaning
..vo- ^^^^^	Clear right hand, press left middle finger
..oo- -....	20

This is analogous to how you count 2 on your right hand because 20 = 2 × 10. Counting by tens, the other numbers up through 90 follow logically:

Notation	Meaning
.ooo- -....	30
oooo- -....	40
....@ -....	50
...o@ -....	60
..oo@ -....	70
.ooo@ -....	80
oooo@ -....	90

And here's a full Chisenbop 99:

Notation	Meaning
oooo@ @oooo	99

If you want to count past 99, you need to remember a number in your head. That number is how many hundreds you need to add to your hand abacus at the end of your calculation. Fortunately, all that involves is tacking the *hundreds* number onto the front of the number shown by your hands. So, if your hands show this:

Notation	Meaning
....@ -oo..	52

and your hundreds number is 3, the number you have counted to is 352, which you can show like this:

Notation	Meaning
....@ -oo..	(3)52

Therefore, to break past 99, clear your hands and say to yourself "one hundred." Both hands will now show zero, to which you will add the hundred in your head by prepending one, and the Hubble horizon is the limit!

Notation	Meaning
....⁻ ⁻....	(1)00
....⁻ -o...	(1)01
....⁻ -oo..	(1)02
....⁻ -ooo.	(1)03
....⁻ -oooo	(1)04
....⁻ @....	(1)05

Addition

Addition is simply counting by larger chunks than 1. For example, to add 15 + 23, place your hands in the 15 position:

Notation	Meaning
...o⁻ @....	15

Now, press two more fingers representing 10 each on the left hand. This means you're adding 20:

Notation	Meaning
.ooo⁻ @....	35

Almost there. Now, press three more fingers on your right hand for the 3 in 23:

Notation	Meaning
.ooo⁻ @ooo.	38

So, 15 + 23 = 38. It might have been easier to do it this way:

Notation	Meaning
..oo⁻ -ooo.	23
.voo⁻ vooo.	(press 15)
.ooo⁻ @ooo.	38

Subtraction

Simply put, if addition is counting forward by chunks, subtraction is counting backward by chunks. Instead of pressing your inner fingers down and moving out, you are lifting your outer fingers and moving in. It's quite simple; it just takes practice.

Here's how to figure 38 – 23 = 15:

Notation	Meaning
.ooo- @ooo.	38
.^^o- @^^^.	(raise 23)
...o- @....	15

Multiplication

Multiplication is just repeated addition. For example, 3 × 15 = 45 means 15 + 15 + 15 = 45:

Notation	Meaning
...o- @....	15
.ooo- -....	30
oooo- @....	45

It's useful to drill in adding by arbitrary chunks, starting with counting by twos, moving up as high as counting by nines, and then moving into counting by double digits. You want your Chisenbop to become an unconscious *motor skill*; otherwise, you might need to keep intermediate results in your head. For example, when multiplying by 15, you might need to repeatedly add 10 and then remember to add 5. Not only is this not the Way of Chisenbop, but it is also error-prone.

Note that when you multiply, you always need to remember a second number in your head, in addition to the *hundreds*—that is, how many times you've added the number by which you're multiplying. (Don't read the result off your fingers until you're done.) So, in the previous example, when you first place your fingers on the table in the 15 position, you say, "One!" When you add a second 15, you say, "Two!" And when you add the final 15, you say, "Three!" *Now* read the result off your fingers: presto, 45!

Division

If multiplication is repeated addition, division is repeated subtraction. You can divide by repeatedly subtracting the divisor until you can't subtract it anymore. The number of times you could subtract the divisor is the result, and the number left over on your fingers is the remainder. Thus, 15 can be subtracted from 45 three times, leaving all fingers clear (a remainder of zero), showing that 45 / 15 = 3, remainder 0. Likewise, 15 can be subtracted from 46 three times, leaving your index finger pressed, showing that 45 / 15 = 3, remainder 1.

In practice, it's probably easier to work in the opposite direction. For example, you can divide a number by 7 by counting sevens until you exceed the number you are dividing. The number of sevens in the number you are dividing is one less than the number at which you finished. For example, to divide 81 by 7, count 7 (1), 14 (2), 21 (3), 28 (4), 35 (5), 42 (6), 49 (7), 56 (8), 63 (9), 70 (10), 77 (11), 84 (12)—oops! That's 12 sevens, which is too many, so there are 11 sevens in 81, with a remainder of 4.

End Notes

1. Lieberthal, Edwin M. 1982. *The Complete Book of Fingermath*. A & W Visual Library.

2. Wikipedia. 2005. "Chisenbop." *http://en.wikipedia.org/wiki/Chisenbop*.

See Also

- Harris, Andy. "Chisenbop Tutorial." *http://www.cs.iupui.edu/~aharris/ chis/chis.html* (this page from Purdue University contains an interactive, animated Chisenbop tutorial and several lessons presented via streaming media).

Count to a Million on Your Fingers

HACK #40

You can use binary arithmetic to count to more than a million on your fingers.

You probably already know that you can use Chisenbop [Hack #39] to count to 100 or more and do simple arithmetic on your fingers. If you switch from the decimal numeral system to binary, however, you can use your fingers to count to about 2^{20}, which is 1,048,576—more than a million![1]

The Binary Numeral System

First, let's review the binary numeral system. (If you don't need a review, you can skip to the next section.)

The number system we use normally is called the *decimal numeral system* because it is based on powers of 10. For instance:

```
4309 = (4 x 1000) + (3 x 100) + (0 x 10) + (9 x 1)
```

Note that in the decimal system there are 10 digits: 0 to 9. Each position in a decimal number corresponds to a power of 10; for instance, 3 is in the hundreds position in the decimal number 4,309.

The *binary numeral system* is based on powers of 2, so there are only two digits, **0** and **1**. These are referred to as *bits*, which is short for *binary digit*. Thus, in binary:

```
10011 = (1 x 16) + (0 x 8) + (0 x 4) + (1 x 2) + (1 x 1) = 19
```

In this hack, I'll highlight binary numbers in bold, so you can tell them apart from ordinary decimal numbers. Without some convention, it would be impossible to tell whether a number like **10011**, with only 1s and 0s, is binary or decimal.

Notice that because only 1 and 0 are used, there's no real multiplication here: we just add up the positions with 1s in them. In the case of **10011**, that's 16 + 2 + 1 = 19.

In Action

Now we can explain how to count on your fingers in binary. The basic idea is to have a finger designate each position. Let's start with one hand. For example, suppose that on the right hand, the thumb is the 1 position, the index finger is the 2 position, the middle finger 4, the ring finger 8, and the pinky 16. Each finger can be down (representing **0**) or up (representing **1**).

Start with all fingers in the down position. Thus, each position has a 0, and this represents the number zero (**00000** = 0). You can represent any 5 bit number with one hand. For instance, to represent **10011** (which is 19 in decimal) using the system outlined in the previous section, place your fingers as in Table 4-4 and Figure 4-2.

Table 4-4. Fingers in the (decimal 19) position

Finger	Value	Gesture	Bit
Pinky	16	Up	1
Ring	8	Down	0
Middle	4	Down	0
Index	2	Up	1
Thumb	1	Up	1

The rule for adding 1 is simple: each time you increment the number, look for the smallest-position finger that is down. Raise it, and lower all fingers in even smaller positions.

For example, starting with 19 as in the previous example, we would raise the middle finger (the 4 position) and lower the index finger and thumb (the 2 and 1 positions, which are those smaller than the 4 position). Keep the pinky and ring fingers in the same position, as shown in Table 4-5 and Figure 4-3. This represents the number **10100** (16 + 4 = 20).

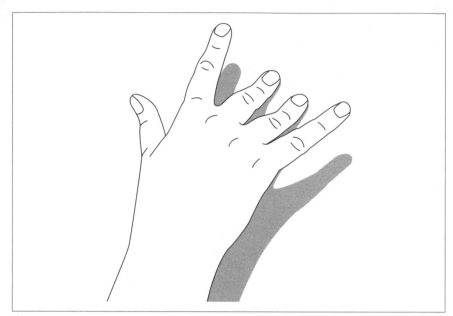

Figure 4-2. Fingers in the 10011 (decimal 19) position

Table 4-5. Fingers in the (decimal 20) position

Finger	Value	Gesture	Bit
Pinky	16	Up	1
Ring	8	Down	0
Middle	4	Up	1
Index	2	Down	0
Thumb	1	Down	0

Incrementing again, we would just raise the thumb (since it's the smallest-position finger that's down). Since there are no smaller positions than the thumb, no fingers are lowered. The number represented is **10101** (16 + 4 + 1 = 21).

With only one hand, we can represent numbers up to **11111** (31). To get up to 1,000, we proceed to the other hand: the thumb on the *left* hand represents the 32 position, the index finger represents 64, the middle finger 128, the ring finger 256, and the pinky 512. With this, we can represent numbers up to **1111111111** (1,023). For instance, 1,000 is represented as **1111101000** because 512 + 256 + 128 + 64 + 32 + 8 = 1,000. (See Table 4-6 and Figure 4-4.)

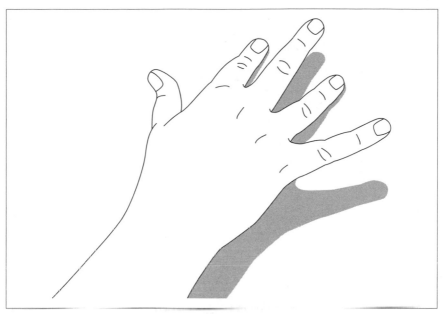

Figure 4-3. Fingers in the 10100 (decimal 20) position

Table 4-6. Fingers in the 11111010000 (decimal 1,000) position

Left hand				Right hand			
Finger	**Value**	**Gesture**	**Bit**	**Finger**	**Value**	**Gesture**	**Bit**
Pinky	512	Up	1	Pinky	16	Down	0
Ring	256	Up	1	Ring	8	Up	1
Middle	128	Up	1	Middle	4	Down	0
Index	64	Up	1	Index	2	Down	0
Thumb	32	Up	1	Thumb	1	Down	0

> An alternate way of figuring this out is to compute the left hand separately, multiply it by 32, and add the right hand. For example, on your left hand: **11111** = 31; 31 × 32 = 992. Then, compute on your right hand: **01000** = 8. Finally, 992 + 8 = 1,000.

That gets us past 1,000, but we're out of fingers. So why did we say you can count to a million this way? The same fingers can *simultaneously* represent a new, *independent* set of bits, which will have values 1,024, 2,048, and so on, up to 524,288. For these higher-valued bits, a finger held straight is off (**0**), and a curled finger is on (**1**).

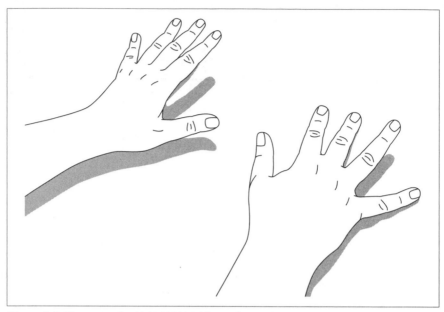

Figure 4-4. Fingers in the 1111101000 (decimal 1,000) position

Now that you've added curled fingers to your binary repertoire, the rule for incrementing must be extended. The previous rule of raising the smallest-position lowered finger and lowering all the fingers in smaller positions still applies. Now, however, if and only if all of your fingers are raised (whether curled or straight), and you want to add 1, lower all your fingers, curl the smallest-position straight finger, and straighten any lower-position curled fingers.

Let's see how we'd represent 1,000,000 with this scheme. 1,000,000 = **11110100001001000000**, because $1,000,000 = 2^{19} + 2^{18} + 2^{17} + 2^{16} + 2^{14} + 2^9 + 2^6 = 524,288 + 262,144 + 131,072 + 65,536 + 16,384 + 512 + 64$.

Splitting this into groups of five (for the five fingers), we must position our fingers as shown in Table 4-7.

Table 4-7. Fingers in the 11110100001001000000 (decimal 1,000,000) position, by hand

Bit	Gesture
11110	Left hand curled/straight
10000	Right hand curled/straight
10010	Left hand up/down
00000	Right hand up/down

Finger by finger, 1,000,000 (**11110100001001000000**) looks like Tables 4-8 and 4-9 and Figure 4-5.

Table 4-8. Lefthand fingers in the 11110100001001000000 (decimal 1,000,000) position, finger by finger

Finger	Value	Gesture	Bit	Value	Gesture	Bit
Pinky	524,288	Curled	1	512	Up	1
Ring	262,144	Curled	1	256	Down	0
Middle	131,072	Curled	1	128	Down	0
Index	65,536	Curled	1	64	Up	1
Thumb	32,768	Straight	0	32	Down	0

Table 4-9. Righthand fingers in the 11110100001001000000 (decimal 1,000,000) position, finger by finger

Finger	Value	Gesture	Bit	Value	Gesture	Bit
Pinky	16,384	Curled	1	16	Down	0
Ring	8,192	Straight	0	8	Down	0
Middle	4,096	Straight	0	4	Down	0
Index	2,048	Straight	0	2	Down	0
Thumb	1,024	Straight	0	1	Down	0

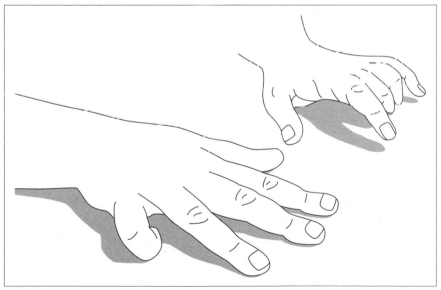

Figure 4-5. Fingers in the 11110100001001000000 (decimal 1,000,000) position

In Real Life

Instead of holding one number on both hands, you can store one number on each hand. This might be useful in keeping score in a two-player game, for instance.

Additional bits are available elsewhere on your body: for instance, your wrists.

Once you get familiar with binary numbers, you can experiment more with addition and even try subtraction (these might make being able to represent numbers up through 1,000,000 more useful).

How It Works

Incrementing is just adding 1, and binary arithmetic works the same as decimal arithmetic (except it's easier because there are only two bits). For instance, the computation we did as the first example (19 + 1 = 20, in decimal) works like this:

```
    11
  10011
+     1
  10100
```

Since the lowest position is **1**, adding **1** gives **2**, which is **10** in binary. So, we put down **0** and carry the **1**. (This corresponds to lowering the thumb.) Adding the carry to the **1** in the 2 position again gives **10**, so we again have **0** with a carried **1**. (This corresponds to lowering the index finger.) In the 4 position, we have a **0**, and **0 + 1 = 1**. (This corresponds to raising the middle finger.) The other digits remain the same, just as the corresponding fingers do.

The next increment (20 + 1 = 21) is simple, because there are no carries:

```
  10100
+     1
  10101
```

This matches what we did with the fingers: we just raised the thumb.

The rule for fingers (find the smallest-position finger that's down, raise it, and lower all smaller-position fingers) corresponds to finding the rightmost **0**, changing it to a **1**, and replacing any **1**s to the right with **0**s. That's exactly what the carries do, as you can see in the first addition.

End Notes

1. Mentat Wiki. "Physio Arithmetics." *http://www.ludism.org/mentat/ PhysioArithmetics*.

—Mark Purtill

Estimate Orders of Magnitude

#41

By using rough order-of-magnitude estimates, you can check calculations and estimate whether tasks are even plausible before spending time to plan them more accurately.

An *order-of-magnitude estimate* is an estimate to the nearest power of 10. For instance, an order-of-magnitude estimate of 400 means the true value is closer to 400 than 40 or 4,000. The estimates we're discussing in this hack are called *rough order-of-magnitude* (ROM) estimates because we're not completely sure we're within an order of magnitude.

Other terms for this kind of estimate are *ballpark estimates* (as in, "Are we even in the ballpark?") and *educated* or *scientific wild-ass guesses*. As all of these names suggest, the idea is to determine roughly how large a number is, or whether a task is trivial, possible, or absurd.

When you're estimating tasks, make sure everyone understands that ROM estimates are just that. They should *not* be written into contracts or used as targets when actually doing work. That needs more careful estimation and project tracking. ROM estimates are just to know whether to bother performing a more formal estimate.

In Action

First, let's do a purely numeric hack: estimate how many seconds are in a year.

There are 60 seconds in a minute and 60 minutes in an hour; multiplying them is easy: 3,600 seconds in an hour. Let's round that to 4,000. Twenty-five percent might seem like a big round up, but we are concerned here with orders of magnitude; as long as it's not 2,500%, we should be OK.

Now, there are 24 hours in a day. We rounded the seconds in an hour up, so let's round this down to 20. Multiplying 20 by 4,000, we get 80,000 seconds in a day, which we can round up to 100,000.

There are 365 days in a year (plus a bit). Let's round that to 400. Our final estimate (400 days times 100,000 seconds) is 40,000,000 seconds in a year, and since we rounded up three times and down once, it's probably high rather than low.

The actual answer, using 365.2422 days per year, is 31,556,926 seconds per year, so we are well within an order of magnitude (and we were indeed high, not low).

Now, let's do a similar problem estimating a task. In an episode of *The Simpsons* I just saw in syndication, Bart is in France as an exchange student for three months, but he ends up picking grapes at a run-down chateau. Supposedly, he picks a million grapes in that time (this is an estimate by one of the Frenchmen exploiting Bart).

First, let's see if 1,000,000 grapes seems plausible. Each bunch of grapes has some grapes on it—more than 10, certainly. Let's say 100. Each vine will have several bunches; let's say 10. So, we've got 1,000 grapes per vine, which means 1,000,000 grapes require 1,000 vines. If anything, that seems low for the vineyard pictured, so no problems here.

Could Bart even theoretically pick all those grapes in less than three months? Let's say Bart picks one grape per second, which seems plausible (he certainly doesn't manage 10 in a second, but he can do more than one every 10 seconds). At this pace, he'd manage 3,600 in an hour. Let's round that down to allow time to move between bushes, empty the container, and so on, but to compensate, we'll have him work 20 hours a day. So, Bart manages 6,000 (20 × 3,000) grapes a day. If we round that to 5,000 to get a nice factor of 10 [Hack #36], we see it will take 200 days (1,000,000 / 5,000, which is the same as 1,000 / 5) to harvest 1,000,000 grapes.

This is somewhat more than three months (which is about 3 × 30 = 90 days), and Bart did other things as well. So, this would seem a bit suspect if it weren't intended as humorous exaggeration—but only a bit suspect, since there were a lot of guesses in there. If this were important, it would be worth doing a more careful analysis of things, such as Bart's grape-picking speed.

How It Works

The basic idea is to estimate anything you don't know and round off numbers to make the arithmetic easy. As you estimate and round, if you keep track of whether you're estimating high (rounding up) or estimating low (rounding down), you get a feeling as to whether your estimate is more likely to be high or low.

Rounding up one time and down the next will tend to cancel out (though of course this depends on how much you're rounding). That's just because using a bigger number in one place will cancel out using a smaller number in another. For instance, when computing the seconds in a day, we rounded 3,600 up to 4,000 and 24 down to 20, and multiplied them to get 80,000. The actual answer is 86,400 (3,600 × 24), and we're quite close. Notice that if we'd rounded both numbers up—say, to 4,000 × 30—we'd get 120,000, which of course is higher, but also further away from 86,400 than 80,000 is.

Likewise, rounding both down—say, to 3,000 × 20—gives us 60,000; again, further away than 80,000.

Sometimes, you want to mostly round up (if it's work you're going to have to do) or mostly round down (if you're estimating feasibility). You might want to do both and use the two numbers as upper and lower bounds. For instance, in the example of the number of seconds in a year, we'd know that the number of seconds in a day is between 60,000 and 120,000.

In Real Life

ROM estimates can help you find mistakes in everyday life. They can help you in your job, and they might even help you get a job if you don't have one.

Checking calculations. Mistakes on calculators can easily give answers that are wildly wrong. For instance, suppose we were computing the number of seconds in a year, and we typed this:

```
60 * 60 + 24 * 365
```

to get this:

```
12360
```

Since our earlier estimate when we tried to calculate the number of seconds in a year was 40,000,000, we know something went wrong.

Work estimates. At my work, we often use ROM estimates to figure out how long a customer request will take. We estimate tasks by week and add them up to get the total time. We can then compare that to the customer's schedule and budget.

Suppose, for instance, that someone asks Ron to write another book of mind performance hacks, but they want it in two months. Can he do it?

The book will comprise about 10 chapters with about 10 hacks in each chapter, for a total of 100 hacks (which might be an overestimate, but it's a nice, round number to shoot for). Some hacks are easier than others are, but we might say that, on average, each takes two hours to research, two hours to write a first draft, an hour for polishing, and an hour for technical review. That's six hours for each hack, for a total of 600 hours. At 40 hours per week, we already have 15 weeks, well more than two months. So, we can already tell that Ron will need some help, and we haven't counted the time to decide which hacks to use.

The task of deciding which hacks to write should also be figured in and is the kind of thing we tend to like to be generous in estimating, because it's hard to quantify. It's much harder to say, as we did with the writing, that it will take a definite amount of time per hack to think them up, and that time will probably have to be spread out more. On the other hand, we have some hacks in this book that might help.

Job interview. Many software companies like to ask mind-expanding questions to prospective hires to test their ability to think big and creatively. One famous example is, "How would you move Mount Fuji?" Let's estimate the number of truckloads of dirt that would generate.

First, how tall is Fujisan? If you know, you can round that off, but I don't. It's a very tall mountain, so it's probably around 10,000 feet. Again, 1,000 is definitely too small and 100,000 is too big. But 20,000 would be equally plausible for this kind of estimate, or even 30,000, though that would make Mount Fuji taller than Mount Everest. If you'd rather work in the metric system, 5,000 meters or 10,000 meters would be a good estimate, which of course doesn't match the estimates in feet.

According to Wikipedia, Mount Fuji is 3,776 meters (12,388 feet) tall. So our 10,000-foot estimate isn't too bad.

Anyway, let's stick with 10,000 feet for now. Mount Fuji appears to be roughly a cone, with the base diameter about twice the height. (If you look at a picture of the mountain, you'll see the base is actually quite broad.) So, if we knew the formula for the volume of a cone, we'd be all set.

First, let's suppose we don't; all we remember is the volume of a cube, which is the side cubed. If we visualize a cube as high as the mountain, it looks *roughly* the same. For a ROM estimate, that's probably good enough, and the cube's volume would be 1,000,000,000,000 cubic feet ($10,000^3 = 10^{12}$).

If you do remember a bit more geometry, you'll recall that the volume of a cone is (area of base) × height / 3, and the area of the base is $\pi \times r^2$. Since we estimated the diameter of the base as twice the height, the radius (r) of the base is the same as the height. Putting this all together, we get $\pi \times$ height^2 × height / 3. And, since π is about 3, we end up with 1,000,000,000,000 cubic feet ($10,000^3 = 10^{12}$) again.

OK, how much can a truck haul away? Probably a volume of 10 × 10 × 10 is not unreasonable (trucks aren't 10 feet wide, but they could be longer and that should make up for the width). So, one truckload is 1,000 cubic feet.

We would need 1,000,000,000,000 / 1,000 = 1,000,000,000 (one U.S. billion) truckloads, which is a lot. In this case, of course, there are many other problems with moving Fujisan; for instance, the Japanese people would probably object!

—Mark Purtill

HACK #42 Estimate Square Roots

Estimate square roots and even higher-order roots by using simple processes.

It's often useful to compute the square root of a number, especially when you want to visualize areas or compute diagonal distances. There are a few methods for computing square roots on paper, some of which are not widely known. If all you have is your brain, however, it's still possible to come up with a quick estimate that's reasonably accurate.

In Action

To estimate the square root of a number, start by pairing up the digits of the number, beginning with the decimal point and moving away. So, for instance, to compute the square root of 500,000, we would pair up the digits like this:

50 00 00

Each pair of digits will, in fact, represent a digit in the square root. The leftmost pair of digits (or single digit, if there are an odd number of digits in your original number) is used to compute the leftmost (most significant) digit of the result. You find the result digit by determining the highest square that will fit into the pair without going over.

The biggest perfect square that fits into 50 is 49, which has a square root of 7. So, we know our square root will have the form 7dd—that is, a 7 followed by two digits, or "seven hundred and something." Further, since 50 is very close to 49, we can surmise that the square root will be in the low 700s. Had it been close to the next square (64, with a square root of 8), we would know that the square root would be closer to 800. Unfortunately, computing the exact value of the later digits requires pencil and paper, but our estimate is in the ballpark.

What about numbers smaller than 1? Basically, the same method applies. You still pair up digits going away from the decimal point, and the most significant digit will still be the square root of the biggest square that fits into the leftmost pair of digits. So, to compute the square root of 0.0038234, you would pair up the digits like so:

```
00 . 00 38 23 40
```

Each pair of digits again represents a single digit in the answer, and the decimal place stays where it is. The biggest square that fits in 38 is 36, with a root of 6. So, your square root will be 0.06.... Since 38 is close to 36, we can also surmise that our approximation is already pretty close to the answer, so the next digit will be small.

Very large and very small numbers. Numbers in science and engineering are often expressed in scientific notation, which is a convenient way to express astronomically large or small numbers. With scientific notation, numbers are expressed in the form:

```
a × 10ᵇ
```

where a is usually in the range 1 <= a < 10, and b is an integer. Nearly all modern calculators use scientific notation for representing large or small numbers; typically, the 10 is omitted and the letter E for "exponent" is used (e.g., 5.2349 E+41). Since the exponent tells us the number of digits, we can use a very similar method to estimate the square root of some very large or very small numbers.

When the exponent b is even, it's easy to come up with an estimate. Simply estimate the square root of the a part and halve the exponent. So, given a number like 5.8345×10^{82}, we can immediately tell that the square root is going to be $2.d \times 10^{41}$, because:

- 4 is the biggest square that fits in 5.
- 4 has a square root of 2.
- The exponent 41 is half of 82.

This works for both positive and negative even exponents.

When the exponent b is odd, we need to borrow the first digit from the other side of the decimal point of a. Estimate the square root of that, and you have the a part of the result. To get the exponent of the result, subtract 1 from the exponent of your original number before halving it.

For instance, given 5.0234×10^{17}, we would find the square root of the biggest square that fits in 50 (49, with a square root of 7) and subtract 1 from the exponent 17 to make it 16 before halving it, giving us a square root of $7.d \times 10^{8}$.

For a number with an odd *negative* exponent, such as 1.9123×10^{-43}, you will still subtract 1 from the exponent before halving (-43 to -44); since 4 is the closest square root to 19, the answer will be $4.d \times 10^{-22}$.

Higher roots. What about cube roots, fourth roots, and so on? There is a more general rule that will let you estimate these. Instead of pairing up the numbers, to compute the nth root of a number, divide its digits into groups of n digits each, going away from the decimal point. So, to compute the fourth root of 7,324,643,245, group it like this:

    73 2464 3245

As before, each group represents one digit of the result. And the first digit will be the number closest to the fourth root of the first group, 73. While there is no easy way to compute or memorize the fourth roots of numbers, remember that it can be only one of nine possible digits, so it is fairly easy to find it via trial and error. $2 \times 2 \times 2 \times 2$ is 16; $3 \times 3 \times 3 \times 3$ is 81, which is a little too large. So, our answer will be 2dd, a number in the high 200s. As you've probably realized, the method for computing square roots described earlier is just a specific ($n = 2$) example of this more general method.

In Real Life

Suppose you come across a real estate listing advertising a house with an area of 3,700 square feet. Breaking this up into pairs of digits gives you:

    37 00

You have two groups, so it'll be a two-digit number. The largest square that fits into 37 is 36, with a square root of 6. And it's pretty close, so this house will have as much floor space as a square house a little longer than 60 feet on a side.

It was recently reported that the ozone hole over Antarctica has shrunk to 6,000,000 square miles. How big is this? Breaking this up into pairs of digits gives you:

    6 00 00 00

The biggest square that fits into 6 is 4 (2×2), so you should be able to determine, with a moment's reflection, that this represents a square area measuring in the middle 2,000s on each side. Still a pretty big hole in the ozone!

If you'd rather picture a circle, it will always have a diameter that's about 13% bigger than the square root you just estimated.

Also in the news at this writing, Hurricane Katrina reportedly left an estimated 16 million cubic yards of debris littering the coastline of Mississippi alone. How big a warehouse would it take to hold all of that? Because it's cubic, we gather the digits in groups of three:

16 000 000

That's three digits, and the cube root of 16 is somewhere between 8 (2 × 2 × 2) and 27 (3 × 3 × 3), so this represents a cube that's about 250 × 250 × 250 *yards*. That's a warehouse wider, longer, and taller than two football fields, and that's just the debris littering Mississippi.

See Also

- Wikipedia. "Square root." *http://en.wikipedia.org/wiki/Square_root*. A good summary of more traditional methods of estimation and exact computation of square roots, including the exact method that this estimation method is based on. Go there to learn how to compute a square root exactly, with the help of pencil and paper.
- Doerfler, Ronald W. *Dead Reckoning: Calculating Without Instruments*. Gulf Publishing Company. Contains further methods for computing roots.

—Mark Schnitzius

 HACK #43 Calculate Any Weekday

Quickly calculate the day of the week for any date of the Gregorian calendar—useful for scheduling appointments and meetings!

The imperfect Gregorian calendar, when combined with the imperfect Earth year—which is an even multiple of neither 12 (the number of months) nor 7 (the number of days in the week), but instead an icky 365.24237404 days long (approximately!)—means that, for most of us to find what day of the week a meeting falls on, we have to consult a wall calendar or our PDA.

But what if you had your own perpetual calendar in your head? What if you could, with practice, take just a few seconds to calculate any day of the week from centuries ago and into the distant future, when they finally nudge the Earth into a more reasonable orbit?

You can. Here's how.[1,2]

In Action

To calculate any weekday, you basically need to find four numbers, add them together, and then *cast out sevens* (i.e., calculate that number *modulo* 7, a simple procedure). In practice, you can do the modulo math as you go along to keep the numbers small and simply keep a running total.

Here are the numbers you need:

- The year-item
- The month-item
- The day-item
- Adjustment

The year-item. Finding the year-item (or *key number* for the year) is easy. Here's the formula:

```
(YY + (YY div 4)) mod 7
```

where YY represents the last two digits of the year.

Modulo Math

The div and mod operators come from integer arithmetic. The div operator is the same as ordinary division, but it discards the remainder. So, 37 div 4 = 9, because the remainder of 1 is discarded.

The mod operator simply finds the remainder when the number before the mod is divided by the number after the mod. So, 37 mod 4 = 1, because 37 / 4 = 9, with a remainder of 1.

Numbers less than 7 (such as 05, the last two digits of the current year 2005) can only be divided by 7 zero times and have themselves as a remainder (so, 5 / 7 = 0r5, 5 div 7 = 0, and 5 mod 7 = 5).

The month-item. Finding the month-item requires some memorization. There are only 12 key numbers, though, one for each month, as shown in Table 4-10.

Table 4-10. Key numbers for calendar months

Month	Key number
January	0
February	3
March	3

Table 4-10. Key numbers for calendar months (continued)

Month	Key number
April	6
May	1
June	4
July	6
August	2
September	5
October	0
November	3
December	5

To memorize these numbers, you can use any mnemonics that you prefer, such as the Dominic System [Hack #6].

The day-item. The day-item is simply the day of the month—for example, 1 for April 1, 31 for October 31, 15 for March 15, and so on.

Adjustment. The fourth number you will need to find is an adjustment to the total. It has two parts: the century-item, plus a possible tweak if the year is a leap year.

Since you'll mostly be finding dates in the 20th and 21st centuries, you can probably ignore most of Table 4-11, and just remember that for dates from 1900–1999, the adjustment is 0 (that is, don't add anything), and for dates from 2000–2099, you add 6.

Table 4-11. Century-item adjustments

Century	Adjustment
1700s	4
1800s	2
1900s	0
2000s	6
2100s	4
2200s	2
2300s	0

A more precise method follows.

The Julian calendar ended in most Western countries on September 2, 1752, and the Gregorian calendar began on September 14, 1752.

They had to fudge the date when they converted over. Legend has it that people rioted for their *lost days*. And you thought Y2K was a big deal.

To get the century-item for any date on the Julian calendar, subtract the century (which would be 14 for the year 1492) from 18, and cast out sevens.

To get the century-item for any Gregorian date, divide the century by 4, cast out sevens, subtract the result from 3, and multiply the difference by 2. Thus:

```
20 / 4     = 5
5 mod 7    = 0
3 - 0      = 3
2 x 3      = 6
```

And 6 is indeed the century-item for the 2000s, as shown in Table 4-11.

Now for the leap-year tweak. If your date is in January or February of a leap year, subtract 1 from the running total.

Any year evenly divisible by 4 in the Gregorian calendar is a leap year, except that years also divisible by 100 aren't—except that years *also* divisible by 400 *are* leap years. Thus, 1936 was a leap year (it is evenly divisible by 4); 1937 was not a leap year (it's not divisible by 4). The years 1800 and 1900 weren't leap years (they're evenly divisible by 100), but 2000 was (it's also evenly divisible by 400).

Leap years in the Julian calendar are simply any year divisible by 4.

What to do with the result. If you have *cast out all sevens* (that is, calculated the number mod 7 correctly), the end result will be a number from 0 to 6. This number will tell you the weekday, starting with Sunday at 0 and counting upward, as shown in Table 4-12.

Table 4-12. Weekdays and corresponding results

Weekday	Result
Sunday	0
Monday	1
Tuesday	2
Wednesday	3
Thursday	4
Friday	5
Saturday	6

In Real Life

Let's calculate the weekday of the first moonwalk, July 21, 1969:

1. The key number for the year is (69 + (69 div 4)) mod 7.

2. 69 div 4 = 17, because 69 / 4 = 17 with a remainder of 1 (or 17r1).

3. 69 + 17 = 86.

4. Now, cast out sevens: 86 mod 7 = 2, because the next highest multiple of 7 is 84, and 86 – 84 = 2. *Remember the number 2.*

5. The key number for the month of July is 6 (see Table 4-10). Add that key number to the result you remembered in the previous step: 6 + 2 = 8.

6. Cast out sevens: 8 mod 7 = 1. (7 goes into 8 once, with 1 left over.) *Remember the number 1.*

7. The key number for the day is 21, because the first moonwalk took place on July 21. Add the result you remembered in the previous step: 21 + 1 = 22.

8. Cast out sevens: 22 mod 7 = 1.

9. The adjustment is 0, because this took place from 1900–1999 and it was not January or February of a leap year: 1 + 0 = 1.

10. The final result is 1, so the first moonwalk took place on a Monday. (And consulting a calendar, we find that this is true!)

To be precise, the moonwalk took place at 2:39:33 A.M. UTC, which was Sunday night throughout the U.S. I bet the workers of the world had quite a bit to talk about around the water cooler the next day.

End Notes

1. Carroll, Lewis. "Lewis Carroll's Algorithm for finding the day of the week for any given date." *http://www.cs.usyd.edu.au/~kev/pp/TUTORIALS/1b/carroll.html.*

2. Mentat Wiki. "Calendar Feat." *http://www.ludism.org/mentat/Calendar Feat* (explains several different ways to calculate the date, and many shortcuts).

See Also

- "Use the Dominic System" [Hack #6] uses memorization of the month-item table as an example.

- If the World Calendar were adopted, the same date would always have the same weekday in every year. Simple, flexible, logical, and utterly unlikely to be adopted, it's the Esperanto [Hack #51] of calendars! (See *http://personal.ecu.edu/mccartyr/world-calendar.html.*)

Decision Making
Hacks 44–49

Whether you're deciding which house or which hamburger to buy, you make decisions every day, if not every minute. It therefore behooves you to learn more about the art and science of decision making.

Because the hacks in this chapter rely on the analysis of future events, they use a lot of math, but don't let that scare you. You can often use the math hacks from Chapter 4 to simplify things.

This chapter addresses the following questions:

- How important is your problem [Hack #44]?
- How long will it last [Hack #45]?
- What steps can you take to solve it [Hack #46] and [Hack #47]?
- What can you do when all analysis fails [Hack #48]?
- Perhaps most importantly, what do you do when you've got that Friday 7:30 feeling in your bones [Hack #49]?

 Foresee Important Problems
HACK #44
Learn to foresee the most significant problems you'll face by multiplying the probability of an event by its impact on human-friendly seven-point scales. The result is a final estimate of importance on a scale of 0 to 100.

This decision-making hack is similar to the technique known as *bullet-proofing*[1] but with a much finer resolution and ability to compare concerns. The idea of bulletproofing is to do "negative brainstorming" about all the things that could possibly go wrong with a project, and then to rank them by priority on a chart labeled "Minor problem" and "Major problem" on one axis, and "Unlikely" and "Very likely" on the other axis. Figure 5-1 ranks four things that could go wrong for an average smoker.

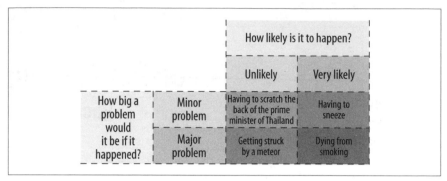

Figure 5-1. A bulletproofing chart for four different problems

In Figure 5-1, "Getting struck by a meteor" and "Dying from smoking" are on the "Major problem" side of the chart, but "Getting struck by a meteor" is on the "Unlikely" side and "Dying from smoking" is on the "Very likely" side. The goal is to attend to the potential problems that are major problems if they occur and are very likely. Sneezing is, of course, very likely, but also not much of a problem; having to scratch the back of the prime minister of Thailand just isn't worth thinking about, because it will probably never happen, and if it did, who cares?

What if we want a resolution that is higher than this simple binary measurement? That's where the Likert Scale comes in. In the early 1930s, the psychologist Rensis Likert (pronounced *lick-ert*) developed the Likert Scale for questionnaires intended to measure attitudes. Attitudes are rated on either a five-point or a seven-point scale; the seven-point scale is considered more accurate, since it has a higher resolution.[2,3]

In Action

For our purposes, the important thing about a seven-point Likert Scale is that it is human-friendly; it makes it easy for humans to convert their fuzzy attitudes, intuitions, and estimates about a phenomenon into a crisp number from one to seven.

A seven-point scale also works well because seven items is about the limit for human short-term memory [Hack #11].

For example, Table 5-1 shows a seven-point Likert Scale designed to capture estimates of probability. Similarly, Table 5-2 shows a seven-point Likert Scale designed to measure how important something would be to someone.

Table 5-1. *Seven-point Likert Scale for probability*

Point	Probability
1	Very improbable
2	Improbable
3	Somewhat improbable
4	Neither probable nor improbable
5	Somewhat probable
6	Probable
7	Very probable

Table 5-2. *Seven-point Likert Scale for importance*

Point	Importance
1	Very unimportant if it happens
2	Unimportant if it happens
3	Somewhat unimportant if it happens
4	Neither important nor unimportant if it happens
5	Somewhat important if it happens
6	Important if it happens
7	Very important if it happens

An interesting thing happens if you combine these scales by multiplication: you get a useful measure of priority. For example, being struck by a meteor might have an importance of 7 (very important if it happens) but a probability of only 1 (very improbable). Thus, it would have a raw priority of 1 × 7 = 7, which should not be considered nearly as high a priority as dying from smoking, with an importance of 7 but also a probability of 7 (very probable), for a raw priority of 7 × 7 = 49. This kind of analysis will help you avoid such common human-reasoning errors as *risk aversion*, where a person will take a guaranteed $40 over $100 with a 50% chance of payoff.[4]

You might also find it interesting that if you double the raw-priority scores, which range from 1 to 49, you get a scale ranging from 2 to 98, which nicely approximates a percentile scale (0–100). There are statistical tricks you can perform to stretch the 2–98 scale onto the percentile scale exactly, but since the math is easy to do in your head without them and nothing is ever utterly important (100) or completely unimportant (0) anyway, why bother?

By the way, if this kind of analysis of concerns and problems depresses you, you can always run the analysis in the other direction and ask what is most likely to go *right*.

In Real Life

Shortly after I developed this hack, I tried it out while designing a new, free (open source) collectible card game called GameFrame (I describe the genesis of this game briefly in "Seed Your Mental Random-Number Generator" [Hack #19]). Some of the concerns I came up with were as follows.

No one will care:

Probability = 4
> Neither probable nor improbable. It's a weird and esoteric game; nevertheless, people do some esoteric things out there on the World Weird Web.

Significance = 6
> Important if it happens. If no one cares, or few people care, it might sink the project.

Priority = 48%
> 4 × 6 × 2

Someone else will publish a similar project first:

Probability = 2
> Improbable. Not only is the game weird and esoteric, but also it's idiosyncratic, which means that if someone else has a similar project, it's likely to be different enough that both projects can coexist.

Significance = 6
> Important if it happens. Even though the chances are low that someone else is producing a nearly identical game, if someone did, it could mean the end of the game project. It's not like a text-editor program, where multiple editors with significantly overlapping functionality can coexist.

Priority = 24%
> 2 × 6 × 2

People will resent the use of free (open source) tools to create the cards:

Probability = 5
> Somewhat probable. People are likely to be ticked off if they have to learn Perl or Scribus to design cards for the card game.

Significance = 3
> Somewhat unimportant if it happens. People can learn, and if they don't want to use free tools, they can produce their own versions of the cards using commercial tools such as Adobe PageMaker.

Priority = 30%
> 5 × 3 × 2

Some people will resent the free (open source) licensing clauses:

Probability = 6
> Probable. People are even more likely to resent the free/open source licensing on the game content than they are on the use of free tools, if they want to make their own proprietary versions of the game, since the free-content rules can't be sidestepped.

Significance = 2
> Unimportant if it happens. That the game will be free and open source is part of the bedrock of the design, so if people don't like it, it's just tough luck; they can go work for Hasbro or something.

Priority = 24%
> $6 \times 2 \times 2$

Someone will claim an intellectual property violation and sue:

Probability = 2
> Improbable. We are being very scrupulous about fair use of copyrighted information and mostly limiting ourselves to original text, graphics, and source code. In any case, people usually have to smell money to sue, and we're a small, shoestring project at the moment.

Significance = 7
> Very important if it happens. No one presently on the project has the time, money, or other resources to defend a case in court. Since the project is primarily a card game and not computer software as such, it's unlikely that the free software community would come to our aid. Also, because of the seemingly frivolous nature of games, we'd be unlikely to see much pro bono help from organizations such as the ACLU.

Priority = 28%
> $2 \times 7 \times 2$

These concerns rank from high to low as shown in Table 5-3.

Table 5-3. Priority for concerns in the GameFrame project

Priority	Outcome
48	No one will care.
30	People will resent the use of free (open source) tools to create the cards.
28	Someone will claim an intellectual property violation and sue.
24	Some people will resent the free (open source) licensing clauses.
24	Someone else will publish a similar project first.

Thus, according to my estimates, the highest-priority concern I should have about the game's success is that no one will care. Accordingly, I made a plan to publicize the game project.

End Notes

1. Bulletproofing. *http://www.mycoted.com/creativity/techniques/bulet-proof.php*.

2. Keegan, Gerard. "Likert Scale" (glossary entry). *http://www.gerardkeegan. co.uk/glossary/gloss_l.htm*.

3. Wikipedia entry. "Likert Scale." *http://en.wikipedia.org/wiki/Likert_scale*.

4. Wikipedia entry. "Risk aversion." *http://en.wikipedia.org/wiki/Risk_aversion*.

HACK #45   Predict the Length of a Lifetime

Many of us instinctively trust that things that have been around a long time are likely to be around a lot longer, and things that haven't, aren't. The formalization of this heuristic is known as Gott's Principle, and the math is easy to do.

Physicist J. Richard Gott III has so far correctly predicted when the Berlin Wall would fall and calculated the duration of 44 Broadway shows.[1] Controversially, he has predicted that the human race will probably exist between 5,100 and 7.8 million more years, but no longer. He argues that this is a good reason to create self-sustaining space colonies: if the human race puts some eggs in other nests, we might extend the life span of our species in case of an asteroid strike or nuclear war on the home planet.[2]

Gott believes that his simple calculations can be extended to almost anything at all, within certain parameters. To predict how long something will be around by using these calculations, all you need to know is how long it *has* been around already.

In Action

Gott bases his calculations on what he calls the Copernican Principle (and what some people call, in this specific application, Gott's Principle). The principle says that when you choose a moment in time to calculate the lifetime of a phenomenon, that moment is probably quite ordinary, not special or privileged, just as Copernicus told us the Earth does not occupy a privileged place in the universe.

It's important to choose subjects at ordinary, unprivileged moments. Biasing your test by choosing subjects that you already believe to be near the beginning or end of their life span—such as the human occupants of a neonatal ward or a nursing home—will yield bad results. Further, Gott's Principle is less useful in situations where actuarial data already exists. Plenty of actuarial data is available on the human life span already, so Gott's Principle is less useful here.

Having chosen a moment, let's examine it. All else being equal, there's a 50% chance the moment is somewhere in the middle 50% of the phenomenon's lifetime, a 60% chance it's in the middle 60%, a 95% chance it's in the middle 95%, and so on. Therefore, there's only a 25% chance that you've chosen a moment in the first fourth of its lifetime, a 20% chance it's in the first fifth, a 2.5% chance it's in the last 2.5% of the subject's lifetime, and so on.

Table 5-4 provides equations for the 50%, 60%, and 95% confidence levels. The variable t_{past} represents how long the object has existed, and t_{future} represents how long it is expected to continue.

Table 5-4. Confidence levels under Gott's Principle

Confidence level	Minimum t_{future}	Maximum t_{future}
50%	$t_{past} / 3$	$3t_{past}$
60%	$t_{past} / 4$	$4t_{past}$
95%	$t_{past} / 39$	$39t_{past}$

Let's look at a simple example. Quick: whose work do you think is more likely to be listened to 50 years from now, Johann Sebastian Bach's or Britney Spears'? Bach's first work was performed around 1705. At the time of this writing, that's 300 years ago. Britney Spears' first album was released in January 1999, about 6.5 years or 79 months ago.

Consulting Table 5-4, for the 60% confidence level, we see that the minimum t_{future} is $t_{past} / 4$, and the maximum is $4t_{past}$. Since t_{past} for Britney's music is 79 months, there is a 60% chance that Britney's music will be heard for between 79 / 4 months and 79 × 4 months longer. In other words, we can be 60% sure that Britney will be a cultural force for somewhere between 19.75 months (1.6 years) and 316 months (26.3 years) from now.

Sixty percent is a good confidence level for quick estimation; not only is it a better-than-even chance, but the factors 1/4 and 4 are easy to use because of the phenomenon of aliquot parts [Hack #36].

By the same token, we can expect people to listen to Bach's music for somewhere between another 300 / 4 and 300 × 4 years at the 60% confidence level, or somewhere between 75 years and 1,200 years from now. Thus, we can predict that there's a good chance that Britney's music will die with her fans, and there's a good chance that Bach will be listened to in the fourth millennium.

How It Works

Suppose we are studying the lifetime of some object that we'll call the *target*. As we've already seen, there's a 60% chance we are somewhere in the middle 60% of the object's lifetime (Figure 5-2).[3]

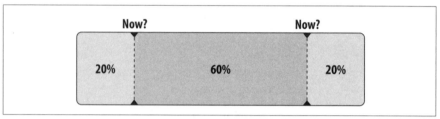

Figure 5-2. The middle 60% of the lifetime

If we are at the very end of this middle 60%, we are at the second point marked "now?" in Figure 5-2. At this point, only 20% of the target's lifetime is remaining (Figure 5-3), which means that t_{future} is equal to one-fourth of t_{past} (80%). This is the minimum remaining lifetime we expect at the 60% confidence level.

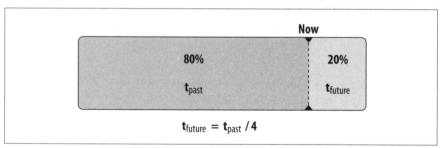

Figure 5-3. The minimum remaining lifetime (60% confidence level)

Similarly, if we are at the beginning of the middle 60% (the first point marked "now?" in Figure 5-2), 80% of the target's existence lies in the future, as depicted in Figure 5-4. Therefore, t_{future} (80%) is equal to 4 × t_{past} (20%). This is the maximum remaining lifetime we expect at the current confidence level.

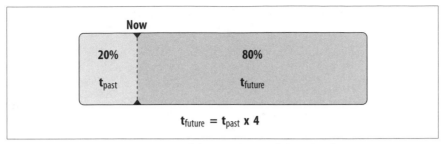

Figure 5-4. The maximum remaining lifetime (60% confidence level)

Since there's a 60% chance we're between these two points, we can calculate with 60% confidence that the future duration of the target (t_{future}) is between $t_{past} / 4$ and $4 \times t_{past}$.

In Real Life

Suppose you want to invest in a company and you want to estimate how long the company will be around to determine whether it's a good investment. You can use Gott's Principle to do so. Although it's not publicly traded, let's take O'Reilly Media, the publisher of this book, as an example.

I certainly didn't pick O'Reilly Media at random, and plenty of historical information is available about how long companies tend to last, but let's try Gott's Principle as a rough-and-ready estimate of O'Reilly's longevity anyway. After all, there's probably good data on the longevity of Broadway shows, but Gott didn't shrink from analyzing them—and I hesitate to say that now that O'Reilly has published *Mind Performance Hacks*, its immortality is assured.

According to the Wikipedia, O'Reilly started in 1978 as a consulting firm doing technical writing. It's July 2005 as I write this, so O'Reilly has existed as a company for approximately 27 years. How long can we expect O'Reilly to continue to exist?

Here's O'Reilly's likely lifetime, calculated at the 50% confidence level:

Minimum
27 / 3 = 9 years (until July 2014)

Maximum
27 × 3 = 81 years (until July 2086)

Here are our expectations at the 60% confidence level:

Minimum

27 / 4 = 6 years and 9 months (until April 2012)

Maximum

27 × 4 = 108 years (until July 2113)

Finally, here's our prediction with 95% confidence:

Minimum

27 / 39 = 0.69 years = about 8 months and 1 week (until mid-March 2006)

Maximum

27 × 39 = 1,053 years (until July 3058)

In the post-dot-com economy, these figures look pretty good. For example, Apple Computer's aren't much better, and Microsoft was founded in 1975, so the same can be said for it. A real investor would want to consider many other factors, such as annual revenue and stock price, but as a first cut, it looks as though O'Reilly Media is at least as likely to outlive a hypothetical investor as to tank in the next decade.

End Notes

1. Ferris, Timothy. "How to Predict Everything." *The New Yorker*, July 12, 1999.

2. Gott, J. Richard III. "Implications of the Copernican Principle for Our Future Prospects." *Nature*, 363, May 27, 1993.

3. Gott, J. Richard III. "A Grim Reckoning." *http://pthbb.org/manual/services/grim*.

HACK #46   Find Dominant Strategies

Sometimes, you can find the best of all possible strategies in what is far from the best of all possible worlds.

Some situations in life are like games, and the mathematical discipline of game theory, which studies game strategies, can be applied to them.

In game theory, a *dominant strategy* is a plan that's better than all the other plans that you can choose, no matter what your opponents do. In other words, a dominant strategy is better than some courses of action in some of the possible situations, and never worse than other courses. Look for a dominant strategy before looking for any other kind of strategy.[1]

In *sequential games*, such as chess or Go, players take turns. You consider your opponent's previous moves, look ahead to anticipate her best moves, and extrapolate to find the optimal move to counter her; the initiative then passes to your opponent, who does the same.

On the other hand, in *simultaneous games*, where players' moves are planned and are executed at the same time, seeking a dominant strategy is helpful. For example, in a presidential debate, you can only guess what your opponent will say and do. In such a situation, using a dominant strategy to know the best possible move regardless of your opponent's move, which you cannot know, is indispensable—if a dominant strategy exists.

In Action

On that world-famous cookery game show, *Titanium Chef*, the contestants are busy cooking on opposite sides of the room, and neither can see what the other is doing. That makes *Titanium Chef* a simultaneous game and an ideal place to look for a dominant strategy.

Consider two contestants, Andi and Bruno. These two chefs must cook in one of two styles: Haute Cuisine and Home Cookin'. Both contestants have made a careful study of the judges' previous preferences, and they know that two of the ten judges prefer Haute Cuisine, and the other eight prefer the guilty pleasure of Home Cookin'.

Furthermore, if Andi cooks in one style and Bruno cooks in the other style, each contestant will get all of the votes from the judges who prefer the particular style. If both contestants cook in the same style, they will split the votes of the judges who prefer that style, and the rest of the judges will pout and abstain. The winner receives $100,000; if there is a tie, the chefs split the prize.

Consider Figure 5-5, which shows the number of votes Andi can expect to get in each possible situation.

		Bruno's Choices	
		Home Cookin'	Haute Cuisine
Andi's Choices	Home Cookin'	4	8
	Haute Cuisine	2	1

Figure 5-5. Possible votes for Andi

If both contestants cook in the Home Cookin' style, they can expect to split the eight votes of the judges who prefer that style, so Andi will get four votes. If both opponents cook Haute Cuisine, they will split the two available votes, and Andi will get one vote.

On the other hand, if Andi cooks Haute Cuisine and Bruno cooks Home Cookin', Andi will get both available votes for Haute Cuisine, for a total of two. If Andi selects Home Cookin' and Bruno chooses Haute Cuisine, Andi will get all eight available votes for Home Cookin'.

No matter what Bruno does, Andi will fare better if she selects Home Cookin', so Home Cookin' is Andi's dominant strategy. You can check this by comparing the top row with the bottom row. Both values in the top row are better than their corresponding values in the bottom row. This means that Home Cookin' *strongly dominates* the Haute Cuisine strategy. If a pair of cells being compared in this case were the same in value, the Home Cookin' strategy would be said to *weakly dominate* the other one.[2]

Now, let's examine Bruno's choices. Figure 5-6 shows Bruno's expected outcome, depending on what each competitor picks.

		Bruno's Choices	
		Home Cookin'	Haute Cuisine
Andi's Choices	Home Cookin'	4	2
	Haute Cuisine	8	1

Figure 5-6. Possible votes for Bruno

Bruno also can expect eight points if he chooses Home Cookin' and Andi chooses Haute Cuisine, two points if the opposite happens, one point if both contestants choose Haute Cuisine, and four points if both contestants choose Home Cookin'.

This time, we're comparing columns, not rows. Both values in the left column (Home Cookin') are bigger than the values in the right column (Haute Cuisine), so Bruno's dominant strategy is also Home Cookin'.

If both players are rational, both will select Home Cookin', since it's the dominant strategy for both of them. If they do so, this episode of *Titanium Chef* will be a foregone conclusion: it will be a 4-4 tie, and each player will receive $50,000.

From Andi's perspective, however, it's always possible that Bruno will miscalculate or not have done his homework and will cook Haute Cuisine instead. In that case, her payoff is huge: she will sweep the judges, receive eight votes, and win $100,000. The same is true from Bruno's perspective.

The worst either of them will do by choosing the dominant strategy is to tie and split the prize, but they have a chance to win outright if their opponent makes a mistake. Without the dominant strategy, they could be the one making the wrong choice, losing outright, and going home with empty pockets.

This simple example is intended for clarity of explanation only. For a more complex example of dominant and dominated strategies, see "Eliminate Dominated Strategies" [Hack #47].

Finding dominant strategies is important because a dominant strategy is your best strategy independent of the impact of your opponents' strategies. It's a way to maximize your potential win and minimize your loss, even before you start and regardless of what happens afterward.

> What is "best" is considered here from a strictly selfish point of view, of course; you might wish to adopt the Golden Rule in some situations despite the fact that it probably wouldn't be a dominant strategy in the game-theory sense.

You can find a dominant strategy in a simultaneous game by creating a table like those shown in Figures 5-5 and 5-6. Populate your table with values calculated any way you prefer, as long as you are consistent. A Likert Scale [Hack #44] provides a human-friendly way of evaluating outcomes.

Note that in our *Titanium Chef* example, both players have a dominant strategy, and it's the same one. Sometimes, however, each player has a different dominant strategy; if that were true on *Titanium Chef*, the contest would still be a foregone conclusion, but there would be a single winner.

Sometimes a player won't have a dominant strategy at all. In that case, he should calculate what the other player's dominant strategy is (if she has one) and make his best response to *that* strategy. It's also important to avoid dominated strategies [Hack #47]. There are also situations (such as a game of Rock Paper Scissors) where no player has a dominant strategy; in such situations, don't overthink things [Hack #48].

In Real Life

Imaginary game shows can be fun, but you might be wondering when you would get a chance to employ dominant strategies in real life. Remember, many real-life situations are gamelike, so you can apply game theory to them. During the Cold War, game theory was even applied to the nuclear arms race, so it can be applied to some very serious "games" indeed. Game theory is also widely used in economics and has even been used to explain some puzzles in evolutionary biology, such as why animals have evolved to cooperate. Consider that all of the following can be modeled as simultaneous games to which game theory can apply:

- Deciding on a legal defense in a courtroom
- Choosing which toys to manufacture for the holiday season
- Deciding whether to attack at dawn
- Deciding whether to be an early adopter of a new technology, or to wait and see if it catches on

As John von Neumann, one of the founders of game theory and inventors of the computer, put it:[3]

> And finally, an event with given external conditions and given actors (assuming absolute free will for the latter) can be seen as a parlor game if one views its effects on the persons involved... There is scarcely a question of daily life in which this problem [of successful strategy] does not play a role.

It is often said that life is a game, but seldom is it said by someone who can back it up with hard figures. Pay attention to dominant strategies, and your life's parlor games may be a little more successful.

End Notes

1. Dixit, Avinash K., and Barry J. Nalebuff 1991. *Thinking Strategically*. W.W. Norton & Company, Inc.

2. Economic Science Laboratory. "Iterated deletion of Dominated strategies." *Economics Handbook*. *http://www.econport.org:8080/econport/request?page=man_gametheory_domstrat*.

3. Bewersdorff, Jörg, translated by David Kramer. 2005. *Luck, Logic, and White Lies: The Mathematics of Games*. A K Peters, Ltd. An excellent recent book on applying game theory to situations people would normally think of as games, such as chess and poker.

Eliminate Dominated Strategies

HACK #47

Find your strongest strategy by systematically eliminating all of your weaker choices.

We've already seen that it's important to find dominant strategies **[Hack #46]** when you make decisions, if possible. If you're lucky enough to have a single dominant strategy, your choice is clear.

Sometimes, however, neither opponent has a dominant strategy. In that case, the opponents should try to eliminate strategies from consideration that are *dominated* and to continue eliminating weaker strategies until a single strategy emerges as clearly superior. When each opponent has settled on a single strategy, they have reached a *pure strategy equilibrium*, which is the best that either opponent can rationally hope for.[1]

In Action

Welcome back to that world-famous cookery game show, *Titanium Chef*. On this episode, we have two time-traveling celebrity chefs named Pasta and Futurio. The ground rules for this episode are as follows:

- Both chefs will choose a cuisine from their respective periods. Pasta will choose between Incan and Sumerian cuisine, and Futurio will choose among Andromedan, Rigelian, and Venusian cooking.

- There are 10 judges on this episode, each of whom may either cast one vote for a chef or abstain from voting.

- Each contestant will take home $10,000 times the number of votes she receives.

The *Titanium Chef* studio has been temporally shielded so that Pasta and Futurio can't use their chronovision sets to predict their opponent's cuisine. However, both Pasta and Futurio do have access to advanced computer simulations that can predict how many votes each chef will receive, depending on which cuisine she chooses. Figure 5-7 shows the possible outcomes in each situation.

Remember, a *dominant strategy* is a plan that's better than all the other plans that you can choose, no matter what your opponents do. In this scenario, you can check whether a strategy is dominant by checking whether all of P's votes in one row are greater than her corresponding votes in another row, or whether all of F's votes in a column are greater than his votes in another column. If one row or column consistently has the most votes, it's a dominant strategy.

		Futurio's Choices		
		Andromedan	Rigelian	Venusian
Pasta's Choices	Incan	P=1 F=3	P=1 F=9	P=6 F=1
	Sumerian	P=3 F=6	P=3 F=1	P=1 F=3

Figure 5-7. All possible outcomes

As you can see, neither opponent has a dominant strategy. For example, Andromedan cuisine does not dominate Rigelian for Futurio, and vice versa, because each is better for one of Pasta's strategies and worse for the other. However, Andromedan cooking *does* dominate Venusian cooking; Andromedan cooking is a better strategy than Venusian whether Pasta cooks Incan or Sumerian.

Thus, since both players are rational, and each knows the other to be rational as well (each knows the other has a reliable simulation of the contest), they both eliminate the *dominated strategy* of Venusian cooking from their calculations, leaving a simplified game that looks like Figure 5-8.

		Futurio's Choices	
		Andromedan	Rigelian
Pasta's Choices	Incan	P=1 F=3	P=1 F=9
	Sumerian	P=3 F=6	P=3 F=1

Figure 5-8. Simplified game without Venusian cuisine

Futurio now has no clear choice, but eliminating Venusian cuisine as an option means that one of Pasta's strategies is now dominated: the Incan row can be eliminated, since both of her values in the Sumerian row are higher. After eliminating the Incan cooking row, the game looks like Figure 5-9.

		Futurio's Choices	
		Andromedan	Rigelian
Pasta's Choices	Sumerian	P=3 F=6	P=3 F=1

Figure 5-9. Simplified game without Incan cuisine

Narrowing Pasta's choices down to Sumerian means that Futurio now has a dominant strategy. Andromedan food clearly dominates Rigelian food, with six votes to Rigelian cuisine's one vote.

Eliminating the dominated Rigelian strategy means that the final pure strategy equilibrium is Sumerian versus Andromedan food. The outcome is that Pasta receives three votes ($30,000) and Futurio receives six votes ($60,000), as shown in Figure 5-10.

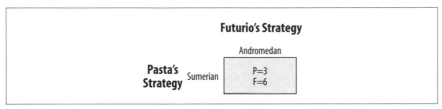

Figure 5-10. The Titanium Chef pure strategy equilibrium

Thus, this episode of *Titanium Chef* was a foregone conclusion, and we didn't even need to watch it or use a time machine to discover how it would turn out. Reality TV tends to work that way...

How It Works

This hack assumes that both opponents are rational. That might seem peculiar; what if your opponent isn't?

It is sometimes possible to do better than game theory predicts, just as it's possible to make a dumb move in chess that *might* pay off, and hope that your opponent doesn't notice how dumb your move was. However, "maybe they won't notice" is not a consistently winning strategy, so the wise player will put up the best possible defense and not pin all his hopes on his opponent being an idiot.[2]

Iterated elimination of dominated strategies works because it simplifies the game in question to a point where it can be handled more easily. Eliminating strategies which neither you nor any rational opponent would play may expose other strategies that can be eliminated the same way. Eventually, either each player will have only one strategy, or the game will at least be simplified to the point that you can analyze it in another way.[3] Think of the process as analogous to reducing a fraction to its lowest terms: the situation being analyzed remains the same, but you can see the answer more clearly.

In Real Life

You can use iterated elimination of dominated strategies in the same real-life situations in which you can find a dominant strategy [Hack #46]. In fact, finding a dominant strategy is just a special case of iterated elimination: in effect, all dominated strategies have already been eliminated.

End Notes

1. Dixit, Avinash K., and Barry J. Nalebuff. 1991. *Thinking Strategically*. W.W. Norton & Company, Inc. The best book of which I'm aware on applying game theory to everyday life.

2. Bewersdorff, Jörg, translated by David Kramer. 2005. *Luck, Logic, and White Lies: The Mathematics of Games*. A K Peters, Ltd.

3. Economic Science Laboratory. "Iterated deletion of Dominated strategies." *Economics Handbook*. http://www.econport.org:8080/econport/request?page=man_gametheory_domstrat.

See Also

- The EconPort digital economics library has an *Economics Handbook* with a wonderfully lucid exposition of basic concepts in game theory: *http://www.econport.org:8080/econport/request?page=man*.

HACK #48   Don't Overthink It

When each side in a game—or an important decision—is trying to outsmart the other, it might be time to flip a coin.

On our third trek through the foothills of game theory, let's leave the wilds of *Titanium Chef* behind. Instead, imagine you are playing a game in which you hold a black Go stone in one hand and a white Go stone in the other. Your opponent must choose the hand holding the white stone. If she chooses correctly, she wins $1 from you; if she does not, she pays you $1.

Now imagine that your opponent is super-intelligent and will always out-guess you. If you intentionally hide the white stone in your right hand, she will choose that hand. If you decide that she knows you will hide the stone in your right hand, and you try to outsmart her and hide it in your left, she will know that you know she knows, and she will decide to pick your left hand. No matter which hand you decide to hide the stone in while trying to outthink her, she will always be able to outthink you and pick the correct hand.

In this situation, the optimal strategy is to shake the Go stones in your cupped hands so that even you do not know which is which, and randomly take one in each hand. In fact, that is always your optimal strategy in this game against a rational opponent, and you must usually assume your opponent is rational. Similarly, your opponent's rational strategy is to flip a coin to determine which hand to pick. You can never expect to do better than 50:50 against a rational opponent in this game, no matter which side you are on, and your optimal strategy is always a randomly chosen 50:50. Therefore, sometimes it's simply best not to overthink things!

In Action

What situations does this hack apply to? Randomness is needed in games when having to go first would be a disadvantage. Consider Rock Paper Scissors (RPS). A player who goes first (this is never supposed to happen in the real game) will always lose to a rational opponent, who will play the perfect countermove: Rock to smash Scissors, Paper to wrap Rock, and Scissors to cut Paper. You can model a super-intelligent opponent simply as someone who always gets to go last.[1]

Extending our reasoning about the hidden-stone game, it's not hard to see that the perfect RPS strategy against a perfect, rational opponent is one-third Rock, one-third Paper, and one-third Scissors, all chosen at random (perhaps by rolling dice). In terms of game theory, this is a *mixed strategy equilibrium*, meaning you are randomly mixing your strategies of Rock, Paper, and Scissors.

"Find Dominant Strategies" **[Hack #46]** and "Eliminate Dominated Strategies" **[Hack #47]** discuss what to do in situations when randomness is not appropriate.

In Real Life

My friends and I play a game called Zendo,[2] in which a player called the Master creates a secret game rule that determines whether any sculpture (called a Koan) made by the other players (called Students) "has the Buddha Nature." As you might guess, the theme of the game is study in a Zen Buddhist monastery. However, a better theme for the game might have been scientific induction, as players attempt to use inductive logic to guess the secret rule. In fact, Zendo is heavily influenced by an earlier game called Eleusis[3] in which the theme of Scientists playing against Nature is made explicit. Both games are excellent training in inductive reasoning skills **[Hack #67]**.

Zendo, and even more so Eleusis, can be summed up by this quote from *Good Omens*, written by Neil Gaiman and Terry Pratchett:

> God does not play dice with the universe; He plays an ineffable game of His own devising, which might be compared, from the perspective of any of the other players, to being involved in an obscure and complex version of poker in a pitch-dark room, with blank cards, for infinite stakes, with a Dealer who won't tell you the rules, and who *smiles all the time*.[4]

When a Student creates a Koan and cries "Mondo!" all Students have the opportunity to guess whether the new Koan has Buddha Nature. If you guess correctly, you get a point, which is good; it lets you attempt to guess the secret rule later.

Interestingly enough, players get to vote on Koans by proffering a Go stone in their closed fist, in a way similar to the guessing game described earlier. My friends and I often find when we are playing the game as Students that in some sense we are trying to outsmart not only the Master, but also *ourselves*; we are often so far from learning the secret rule that whatever we guess consciously is comically bound to be wrong. In this case, we shake two Go stones in our cupped hands and proffer whichever one ends up in our right hand. This strategy tends to maximize our profit: the Master gives us points 50% of the time (more or less), we still learn something, and we have enough points when we need them to guess the secret rule.

Just so, sometimes in that "real" game against Nature, you have to trust and let go. Talk to a stranger. Delurk on a mailing list. Put up a blog or a wiki and see who stops by. Sometimes it's spammers, sometimes it's cranks, and sometimes it's someone you really want to know. A random play in the game of life [Hack #49] can sometimes have a big payoff.

End Notes

1. Dixit, Avinash K., and Barry J. Nalebuff. 1991. *Thinking Strategically*. W.W. Norton & Company, Inc.

2. Heath, Kory. 2003. "Zendo." *http://www.wunderland.com/WTS/Kory/Games/Zendo*.

3. Abbott, Robert. "Eleusis." *http://logicmazes.com/games/eleusis.html*.

4. Gaiman, Neil, and Terry Pratchett. 1992. *Good Omens: the Nice and Accurate Prophecies of Agnes Nutter, Witch: a Novel*. Berkley Books.

See Also

- Bewersdorff, Jörg, translated by David Kramer. 2005. *Luck, Logic, and White Lies: The Mathematics of Games*. A K Peters, Ltd.

Roll the Dice

HACK #49

Break out of your rut by making lists of tasks, recreations, books to read, or research avenues to investigate—including some you don't want to—and rolling the dice to determine your fate.

Dicing, or *dice living*, is a decision-making technique developed by "Dice Man" Luke Rhinehart (pen name for George Cockcroft). While dicing is not a panacea, it is a many-sided remedy. It can:

- Break through your analysis paralysis
- Bring more fun into your life
- Introduce novelty and unpredictability into the way you do things

As Rhinehart writes:[1]

> Dicing is simply one of many ways to attack seriousness. If you list six options, some moral, some immoral, some ambitious and some trivial, some spiritual and some lusty, and let chance decide what you do, then you are in effect challenging the seriousness of your acts, you are saying it doesn't matter what I do. When the die chooses an action I choose to do it with all my heart—that is the dice-person's controlled folly.

Controlled folly is a term that Rhinehart appropriated from the fiction of Carlos Castaneda. To act with controlled folly is to act with the belief that your actions are useless, but to do them anyway, and to care about them. This is the essence of dicing.

In Action

The next time you're bored or depressed, make a list of six or more possible options (for example, reading a random book, getting drunk, having sex, writing a book, working overtime, going to the gym). Then, roll a die or dice to choose among them.

> It's important to include on your list some options that are distasteful to you, some that are boring, and some that are frightening (skydiving, for example), at least from time to time. Part of the value of using dice to decide is to create the possibility of shaking up your life when you shake the dice.

Whichever option comes up, it's crucial to the hack that you be completely obedient to the dice roll and do what it "tells" you to do. Otherwise, why roll the dice in the first place, except to gain insight into what you *really* want?

Dicing can also involve randomly role-playing [Hack #31] characters, emotions, and relationships with other people (patient/therapist, parent/child, lovers, enemies, and so on), always with *controlled folly*.

The Code

You can use the *pyro* program [Hack #20] to generate random options for you even more flexibly and powerfully than dice will.

Since my wife and I live in a Seattle suburb, we use *pyro* to generate things to do in the Seattle area. Here's our datafile, *whattodo.dat*:

```
#@activity@
@funout@
@funout@
Order out from @takeout@. @funhome@
Stay home. @funhome@

#@funhome@
Cook a tasty meal together.
Design a game.
Go catalog shopping.
Just talk together.
Listen to @webradio@.
Listen to some audio together (roll for which).
Make jewelry together.
Play @game@.
Randomly surf the Web via @randomweb@.
Read aloud.
Reorganize @mess@.
Watch a movie at home (make a list and roll).
Work on a self-help book together.
Work on the Glass Bead Game together.
Write some parody lyrics.

#@funout@
Drive randomly, starting in @watown@.
Drive through downtown Seattle looking for adventure.
Eat out; make a list of restaurants and roll.
Go (window) shopping at @store@.
Go letterboxing.
Go on a thrift store expedition.
Go people-watching at @park@.
Go to Marymoor Park with the dogs.
See a movie (make a list of current movies and roll).
Take a road trip to Portland.
Visit a cafe.
Visit the @museumetc@.
Visit the library.
```

#@game@
Blokus
Boggle
Can't Stop
Carcassonne
Cathedral
Focus
Pickomino
Ploy
Rummy
Scrabble
Ultima

#@mess@
the bedroom
the computer room
the garage

#@museumetc@
Asian Art Museum
Experience Music Project
Science Fiction Museum
Seattle Aquarium
Seattle Art Museum
Woodland Park Zoo

#@park@
Occidental Park
the Des Moines pier

#@randomweb@
RandomURL.com
random Wikipedia pages
random pages from H2G2

#@store@
@bookstore@
@gamestore@
Fry's
Ikea
Math'n'Stuff
Silver Platters

#@bookstore@
Elliott Bay
Half Price Books
Third Place Books
Twice-Sold Tales
some chain bookstore that begins with 'B'
the University Bookstore

```
#@gamestore@
Game Wizard
Gary's Games and Hobbies
Genesis Games and Gizmos
Uncle's Games

#@takeout@
Chopsticks
Golden Dynasty
Jet City Pizza
Longhorn Barbecue

#@watown@
Bellevue
Redmond
Renton
Seattle
Tacoma
Tukwila

#@webradio@
BBC7
KEXP
This American Life
```

If you use this data, customize it with your own interests and locales; otherwise, it will be of no use to you, except to spy on what a tame life I lead in Seattle.

Running the Hack

See the "How to Run the Programming Hacks" section of the Preface if you need general instructions on running Perl scripts. If Perl is installed on your system, save the *pyro* script and the *whattodo.dat* file in the same directory, and then run *pyro* by typing the following command within that directory:

```
perl pyro whattodo.dat activity
```

If you're on a Linux or Unix system, you might also be able to use the following shortcut:

```
./pyro whattodo.dat activity
```

The following console session shows the generation of weekend plans for Friday, Saturday, and Sunday nights in the Seattle area:

```
$ ./pyro whattodo.dat activity
Order out from Jet City Pizza. Listen to BBC7.
$ ./pyro whattodo.dat activity
Order out from Golden Dynasty. Design a game.
$ ./pyro whattodo.dat activity
See a movie (make a list of current movies and roll).
```

The key to "dicing up" activities with *pyro* is not to generate a huge list of things to do, then pick and choose (as you might when you morphologically force connections [Hack #20]), but to generate *one* activity for an evening and choose to do whatever comes up with the whole of your heart.

End Notes

1. Rhinehart, Luke. 2000. *The Book of the Die: A Handbook of Dice Living*. The Overlook Press.

See Also

- Luke Rhinehart's novel *The Dice Man* (1971) is the book that started it all, but *The Book of the Die* (2000) is a fun tome in the style of an oracle like *I Ching*.

CHAPTER SIX

Communication
Hacks 50–56

Chapter 3 explores the origin of ideas, and Chapters 1 and 2 explore their storage and retrieval. In between, however, there is another step: communication. If an idea in your brain did not originate there, it sprang from the creative act of another person, was imparted to you via communication, and only then was stored in your brain via memory.

This chapter is all about communicating clearly, cryptically, creatively, and in other ways. Whether you're interested in getting your point across, concealing information from your enemies, or thinking and expressing yourself in novel ways, this chapter has hacks for you.

HACK #50 Put Your Words in the Blender

It may seem counterintuitive, but you can be more expressive if you squish and mangle your language.

James Joyce wrote his last book, *Finnegans Wake*, in a language for the third millennium, a language of dreams. He called it *nat language*, a phrase that blends *night language* and *not language*. I call it Blǝnder, a blend of *blender* and *blunder*, because it mixes up words and people sometimes speak it by mistake.

> Phonologists term that upside-down *e* in Blǝnder a *schwa*, and it's pronounced "uh."

Consider Lewis Carroll's (another master of Blǝnder) description of portmanteau words:

> Take the two words "fuming" and "furious." Make up your mind that you will say both words, but leave it unsettled which you will say first. Now open your mouth and speak. If your thoughts incline ever so little towards "fuming," you

will say "fuming-furious;" if they turn, by even a hair's breadth, towards "furious" you will say "furious-fuming;" but if you have the rarest of gifts, a perfectly balanced mind, you will say "frumious."[1]

Blənder cannot represent reality perfectly, but it approximates it more closely than ordinary language does. Think of it as linguistic antialiasing or, to mix a metaphor (since we're mixing everything else), a linguistic triangulation on the object of discussion, pinpointing it along a fuzzy word spectrum [Hack #34].

Paradoxically, by blurring the boundaries between words, a portmanteau such as *frumious* is closer to a certain state of mind (frumiousness) than ordinary English words (*furious*, *fuming*) can ever be.

In Action

Most people find *Finnegans Wake* hard to understand; some consider it mere nonsense. If you examine it closely, however, it uses a form of seman tic data compression. Consider the following sentence from the book:

> When a part so ptee does duty for the holos, we soon grow to use of an all-forabit.[2]

You can interpret this in a number of ways, all of which can be correct. *Ptee* means *petit*, French for *small*, but also *p-t*, the letters of the alphabet. *All-forabit* is a blend of *alphabet* and *all for a bit*. When we understand that *holos* is Greek for *whole*, we are ready to interpret this sentence as something like this:

> When such a small unit as a letter or word does duty for entire objects, we soon grow to use, and grow used to, an alphabet. When small parts do duty for the whole, we soon grow to use (and grow used to) representing everything at once by a small part.

This sentence is Joyce's critique of language: we cannot represent reality accurately, because any use of *p* and *t* is bound to reflect only a fragment of it. It is also his praise of language (we can do so much with so little), and his description of how *nat language* or Blənder works.

The passage is also a good example of the seeming prescience of the *Wake*. I'm not going to claim that *Finnegans Wake* has magic powers, but I will point out Joyce's use of seemingly up-to-the-moment words such as *bit* (as in 0 or 1, the basic unit of information) and *holos*, which could be taken as the plural of *holo* or *hologram*. The fact that these words, and such postmodern concerns as the relationship of part to whole and information to reality, appear in a book written so long ago suggests to me that it still has much to offer in the 21st century.

How It Works

Paronyms are words that are spelled or pronounced somewhat differently from each other and have somewhat different meanings. If you combine two paronyms, you can create a *blend*—or as Carroll called it, a *portmanteau*.

Some research has been done on paronymic blending by psychologists studying production errors in speech, such as *malapropisms* (speech errors involving paronymic substitution, such as "Stop that! You're making a speculum of yourself!"). Malapropisms are common among words that sound similar, have the same number of syllables, and are stressed identically.[3]

This suggests that phonologically similar words are stored near one another in the human neural lexicon. Researchers have suggested that blending "errors" occur when two words in the lexicon, such as *imposter* and *impersonator*, are similar semantically and enter into the later stages of sentence production together, forming a portmanteau like *imposinator*.[4] Since semantically and phonologically similar words seem to be stored near one another in the brain, it might be possible to learn to communicate fluently in Blənder.

But what if people don't understand the blends? Certainly, ordinary static language might be needed for technical communication. In other situations, people could blend at will, and explain their blends in ordinary language if necessary. Successful blends might be reused, and the vocabulary of Blənder would grow.

Eventually, Blənder might be able to describe in a word a freeform emotion, state of mind, or other complicated phenomenon that would require a sentence of ordinary language. Dictionaries of neologisms and foreign words (not necessarily blends), such as *They Have a Word for It*,[5] *The Deeper Meaning of Liff*,[6] and *Family Words*,[7] do this already

Why not increase the expressiveness of our language, by using and spreading these and other new words?

In Real Life

Here are a few Blənders I've collected over the years. *Swet dreams* was produced with a "perfectly balanced mind" when wishing someone goodnight, and I actually believed when I was a boy that *pless* was a common word.

Fashist
> Fashion fascist: one who aggressively promotes his own style of dress or anything else

Pless
 To press something into place

Sedase
 To seduce by sedation, or sedate by seduction

Swet dreams
 Sweet, wet, sweaty dreams

Blander is not the province of only great writers of literature such as Joyce and Carroll. Everyone can use it to blur his language and at the same time enhance its precision. One English translation of the Joycean phrase *nat language* is *night language*, and *Finnegans Wake* depicts an epic dream, so Joyce might have used a tool such as onar [Hack #28] to learn how to blend fluently. It worked for me.

So try onar. Keep a dream diary [Hack #29]. Read *Finnegans Wake*. Practice punning. Blend a few paronyms. A new form of language might be near.

End Notes

1. Carroll, Lewis. *The Hunting of the Snark*. Project Gutenberg. *http://www.gutenberg.org/dirs/etext91/snark12h.htm*.

2. Joyce, James. 1939. *Finnegans Wake*. *http://www.trentu.ca/jjoyce/fw-19.htm*.

3. Fay, David, and Anne Cutler. 1977. "Malapropisms and the structure of the mental lexicon." *Linguistic Inquiry*, 8(3): 505–520.

4. Foss, D. J., and D. T. Hakes. 1978. *Psycholinguistics: An Introduction to the Psychology of Language*. Prentice Hall.

5. Rheingold, Howard. 2000. *They Have a Word for It: A Lighthearted Lexicon of Untranslatable Words and Phrases*. Sarabande.

6. Adams, Douglas, and John Lloyd. 1990. *The Deeper Meaning of Liff*. Three Rivers Press.

7. Dickson, Paul. 1998. *Family Words: The Dictionary for People Who Don't Know a Frone from a Brinkle*. Broadcast Interview Source, Inc.

See Also

- Some fans of *Finnegans Wake* tend to praise it excessively. (You might even think I'm one of them.) As an antidote, Stanislaw Lem has provided a marvelous satire in his collection *A Perfect Vacuum* (Harcourt Brace Jovanovich). The *Wake* compacts all of history into the single nightlong dream of an innkeeper; Lem's imaginary book *Gigamesh* compresses it into the 36 minutes it takes a condemned man to walk to the gallows.

- Allan Metcalf has an excellent book called *Predicting New Words: The Secrets of Their Success* (Houghton Mifflin Company). He writes, "Whether a new word survives does not depend on whether it fills a perceived gap... [nor] on whether it is useful... [but] rather on whether the word resonates with the speakers of a language, and that depends on a number of factors..." He spends much of the rest of the book exploring those factors.

HACK #51 Learn an Artificial Language

Since the language you speak influences your thoughts, speak some unusual languages and open your mind to unusual thoughts.

A *conlang* is a *constructed language*, more commonly known as an *artificial language*. Unlike C and Java, which are artificial languages created by humans for computers to use, conlangs are created by humans for humans to use.

Over the centuries, hundreds of conlangs have been created, and certainly many more projects that are private have never been published. Many conlangs were created with a specific purpose in mind, such as J.R.R. Tolkien's Elvish languages Sindarin and Quenya, which were created for their sheer beauty. Others were created for a laugh or to be used in movies or books.

Some languages are working languages, however, designed to liberate and empower human minds in a specific way. Esperanto, for example, was designed to break down cultural barriers between people of different cultures, and Lojban was designed to remove as many limitations on human thought as possible.

In Action

Here are six well-known constructed languages that can help you think and express yourself in novel ways.

Esperanto. Esperanto is the most widely spoken conlang on Earth, with an estimated 2 million speakers, putting it on par with Lithuanian, Icelandic, and Hebrew.[1] It was designed in 1887 by Dr. L.L. Zamenhof as a kind of neutral, universal second language that would allow native speakers of all languages to meet one another on even ground, with none having an intrinsic fluency advantage.

Esperanto is extremely simple, regular, and easy to learn. It's also extremely flexible. To quote Esperantist Ken Caviness, "It's been used in all conceivable circumstances for over 100 years. Whatever you have to say, you can say it in Esperanto."

One common complaint about the vocabulary of Esperanto is that it is too Indo-European, and most Esperanto words do indeed come from West European languages. However, Esperanto's agglutinative grammar is more akin to other language families. In any case, Esperanto speakers come from all over the world—it's especially popular in China—and I have had mind-opening, preconception-destroying conversations on many subjects with people from many lands in Esperanto.

Esperanto has a wealth of translated world literature, and it can literally open doors for you with its *Pasporta Servo*, an amazing international hospitality service of friendly people in many different countries who make free lodging available for traveling Esperantists. A wide variety of Esperanto learning materials is available on the Web, as are volunteers who will teach it to you free of charge via email. If you're waiting for an engraved invitation to the world, one of those could probably be arranged, too—with an Esperanto postage stamp.

Lojban. Lojban is an elaborate constructed language that was designed to test the Sapir-Whorf hypothesis (see the "How It Works" section of this hack). It's the more robust descendant of the original Loglan project, which was designed by Dr. James Cook Brown. The project forked because of an intellectual property dispute; you might say Lojban is to Loglan as GNU/Linux is to Unix.

Lojban is designed to remove as many restrictions as possible on "creative and clear thought and communication." To this end, its grammar is based on propositional logic, and it has a culturally neutral vocabulary that was algorithmically derived from the six human languages with the most speakers (Mandarin Chinese, English, Hindi, Spanish, Russian, and Arabic).

Lojban's grammar is more regular than even Esperanto's, so much so that it can be fully specified on a computer with a program such as YACC. This highly regular grammar leads to Lojban's famous *audiovisual isomorphism*, meaning that spoken Lojban can be unambiguously transcribed; you even pronounce punctuation. Lojbanists speculate that this feature might be useful for human-computer communication. Science fiction has beaten them to it, however; the characters in Robert Heinlein's 1966 novel *The Moon Is a Harsh Mistress* use Loglan for just that purpose.

In short, Lojban is a kind of super-language designed to shoot off your Sapir-Whorfean linguistic shackles and blow your mind open. Give it a try.

Klingon. Klingon is another popular conlang. Professional linguist Marc Okrand developed it for the *Star Trek* movies, as the language of the Klingons, an alien race.

Dr. Okrand explicitly designed Klingon to be alien, to stretch the human brain by violating human linguistic universals. For example, its syntax uses a word order seldom observed among human languages. Dr. Okrand has a puckish sense of humor and has added other features that are hard for humans to wrap their minds, lips, and larynxes around, but despite this, Klingon has a devoted fan base.

Here's an example of how Mark "Captain Krankor" Mandel, Chief Grammarian of the Klingon Language Institute, hacked his mind with Klingon. You can, too!

> Mandel offers a story about the way Klingon makes him feel. During one of the annual *qep'a'*, he went out on a mission to pick something up for the convention. A light rain was falling; he felt wet, tired, and a little droopy. But he strengthened his resolve by saying a Klingon phrase to himself: *jISaHqo'*, which he says could be translated as "I refuse to care," "I *will not* care," and "I refuse to let this bother me." In English he would only have had the weaker phrase, "I don't care," which wouldn't have conveyed the strength of his intentions. He smiles at the memory. Thinking in Klingon reminded him that being irritated by a little rain was the sort of thing only a foolish human would do.[2]

AllNoun. More a constructed *grammar* than a constructed language, Tom Breton's AllNoun has a vocabulary and a grammar that consist entirely of English nouns, thus embodying an idea first proposed in Jack Vance's 1958 science fiction novel, *The Languages of Pao*.

An AllNoun sentence is a web of relationships with a weirdly static, timeless feel. Here's an example of a sentence written in AllNoun:

> act-of-throwing:whole Joe:agent ball:patient

And here's a rough transliteration:

> In some context, there is an act of throwing, and the agent of that act is Joe, and the patient is some ball.

which conveys this basic intended meaning:

> Joe throws the ball.

You might think that *act-of-throwing* is an attempt to smuggle a verb into the sentence, but in a full constructed language, as opposed to the prototype project that AllNoun is, that hyphenated word would be a timeless, tenseless noun in its own right.

With only one part of speech, AllNoun's grammar is extremely simple. Paradoxically, if you try hacking your mind with AllNoun, you might find its simplicity to be the most difficult, yet most rewarding, aspect of this language.

Solresol. Solresol was developed in 1817 by Jean François Sudre. It was the first international auxiliary language comparable to Esperanto to receive serious attention. Its most salient feature is that it is composed entirely of musical notes. For example, its name, which means simply *language*, consists of the notes *sol-re-sol* from the Western *solfege* scale (*do, re, mi, fa, sol, la, ti (or si), do*).[3]

Although the principle of forming antonyms by reversing the notes in a word is interesting (for example, *fasimisi* means *advance* and *simisifa* means *retreat*), there's probably not much in Solresol to broaden your mental horizons.

Its real value instead comes from enabling you to communicate multimodally. You can express Solresol syllable-notes via singing, playing a musical instrument, flashes of light, semaphore, spoken language, written language, musical notation, and so on. You can even use it to add another information "channel" for modifying the meaning of verbal language.

How It Works

The controversial Sapir-Whorf hypothesis[4] states that there is a direct relationship between the categories that are available in a language and the way the speakers of that language think and act. While if it were true that language completely determined thought, people would never have any thoughts that they could not express (and we know that's not true), there have nevertheless been some suggestive experiments in this area. For example, it was recently shown that one Amazonian tribe without words for numbers greater than two cannot count reliably higher than two or three.[5] Skeptics have pointed out that causation may run the other way: the tribe never developed words for numbers because they never needed to develop the concepts.[6]

As mentioned earlier, Lojban was designed to be a comprehensive test of the Sapir-Whorf hypothesis that enabled its speakers to think logically, clearly, and creatively. However, it's doubtful that there will ever be a full-scale experiment of the sort its designers envisioned, because there may never be many fluent speakers of Lojban, and the number of native speakers is approximately none.

Design Your Own

You can learn a lot about language and the human mind by creating your own language. Does that sound audacious? The Language Construction Kit web site[7] by Mark Rosenfelder explains how to create your own language sounds, alphabets, words, grammar, even speaking and writing style, as well as an imaginary history for your language, and a family of related languages.

Here are a few tips for doing so, partially inspired by Newitz and Palmer's flowchart in *The Believer*[8] but mostly indebted to the Language Construction Kit. For a crash course in linguistics and cross-cultural human thought styles, and much more information than I can possibly convey here, please visit that site.

Models. First, you'll need a basic model and approach for designing your new language:

- Decide whether you want your language to be "natural" (full of irregularities, like English), or "unnatural" (simple and logical, like Esperanto and Lojban).
- Steal from some languages very different from English, such as Quechua, Swahili, and Turkish.

Sounds. Give some thought to how you want your language to sound when spoken:

- Learn something about the disciplines of phonetics and phonology, so as not to make newbie mistakes with the sounds of your language.
- Learn about how vowels and consonants work, including how to invent new ones.
- Decide how your language stresses words.
- Decide whether your language uses tones (like Mandarin Chinese) and, if so, how they work.
- Design your conlang's phonological constraints. For example, *tkivb* could never be an English word, but it might be OK in another language.
- Are you designing a language for aliens? If so, make sure your conlang sounds *really* weird.

Alphabets. Every written language needs an alphabet, the basic building blocks of words:

- Develop a Roman *orthography*—that is, a way of spelling your language with the Roman alphabet.
- If appropriate (for example, for a fantasy language), develop a new alphabet, too.
- Use diacritics and accent marks if you want, but not haphazardly.
- Alternatively, invent pictograms, logograms, a syllabary, or some other way of writing your language.

Word building. Once you've got your alphabet, consider how the letters will be used to form words:

- Determine whether you want a small or a large vocabulary.
- Develop a vocabulary that respects your conlang's phonological constraints. You can write a computer program to generate random words within those constraints.
- Determine whether you want to borrow words from other languages a little, a lot, or not at all.
- Decide whether your language uses onomatopoeia and other sound symbolism (e.g., *buzz*, *tinkle*, *gong*, *rumble*, etc.).
- Be careful not to just reinvent English idioms.

Grammar. Grammar is one of the most complex aspects of a conlang, and even the Language Construction Kit doesn't address it fully. Here are a few issues you'll have to face:

- Determine whether you have nouns, verbs, and adjectives. (Lojban makes do with one part of speech for all three, and adverbs, too.)
- Determine your pronouns: I, you, he, she, it, and many more are possible.
- Determine the order in which parts of speech appear in sentences (in English we use SVO, or subject verb object).

Style. Give your language a distinct personality and feel:

- How is politeness expressed in your language?
- What forms does poetry use in your language? Rhyme and meter? Alliteration, as in Old English? Counting syllables, as in Japanese haiku?

Language families. Assign your language to a group of speakers:

- If your language belongs to an imaginary people, is it derived from other imaginary languages? Tolkien's languages were related in a huge imaginary tree.
- Does your language have dialects?

Speaking and writing. Once you've created your language and formed a group, get communicating!

End Notes

1. *Esperanto: Frequently Asked Questions*. 1999. "How many people speak Esperanto?" *http://www.esperanto.net/veb/faq-5.html*.

2. Newitz, Annalee. 2005. "The Conlangers' Art." *The Believer*, May 2005. *http://www.believermag.com/issues/200505*.

3. Wikipedia. 2005. "Solfege." *http://en.wikipedia.org/wiki/Solfege*.

4. Wikipedia. 2005. "Sapir-Whorf hypothesis." *http://en.wikipedia.org/wiki/Sapir-Whorf_Hypothesis*.

5. Gordon, P. 2004. "Numerical Cognition Without Words: Evidence from Amazonia." *Science*, 306: 496–499. *http://faculty.tc.columbia.edu/upload/pg328/GordonSciencePub.pdf*.

6. Gordon, P. 2005. "Author's Response to 'Crying Whorf'." *Science*, 307: 1722.

7. Rosenfelder, Mark. 2005. "The Language Construction Kit." *http://www.zompist.com/kit.html*.

8. Newitz, Annalee, and Chris Palmer. 2005. "Build Your Own Conlang." *The Believer*, May 2005. *http://www.believermag.com/issues/200505*.

See Also

- Communicate less judgmentally and more dispassionately with E-Prime [Hack #52].

- Write faster with Dutton Speedwords [Hack #14].

- Expand your idea space [Hack #34] and express fine shades of meaning not present in English with neologisms [Hack #50].

- Learn more about Esperanto at *http://www.esperanto.net*.

- Learn more about Lojban at *http://www.lojban.org*.

- Learn more about Loglan at *http://www.loglan.org*.

- Learn more about Klingon at *http://www.kli.org*.

- Learn more about AllNoun at *http://www.panix.com/~tehom/allnoun/allnoun.htm*.

- Learn more about Solresol at *http://www.ptialaska.net/~srice/solresol/intro.htm*.

Communicate in E-Prime
#52

Eliminate the verb "to be" from your communication to become less dogmatic. Easy to learn, hard to master.

Although almost anyone can speak dogmatically *without* using the verb "to be" and also speak calmly and rationally *with* it, many people have found that eliminating "to be" from their speech and writing makes it easier to communicate flexibly and nondogmatically.[1]

Alfred Korzybski, the founder of the discipline of general semantics, thought that the verb "to be" could lead to confused thought, confused action, and even fascism. Because so much of fascism consists of vilification of the enemy, and because so much of vilification of the enemy consists of calling them subhuman, identifying them with the forces of evil, and so on, fascists might find it hard to write propaganda without "to be."

Korzybski considered the use of "to be" as an auxiliary verb ("I am going next door") fairly innocuous, as well as several other uses. He primarily objected to two uses of the verb "to be," which he called *identity* and *predication*.

An example of identity:

Identity
"That is a spaceship!" (Implication: call out the National Guard! Get the White House on Line 1!)

E-Prime alternative
"That certainly looks like a spaceship to me." (Implication: it merits further investigation. What does it look like to you?)

An example of predication:

Predication
"Fred is disgusting." (Implication: ostracize him!)

E-Prime alternative
"I don't like Fred; I find him disgusting." (Implication: but maybe you like him. What do you like about him? Let's talk.)

Of course, "I find Fred disgusting" contains an implicit form of the verb "to be":

I find Fred *to be disgusting*.

Readers might raise other objections. For example, Rational Emotive Behavior Therapy [Hack #57] questions the utility of "rating" human beings at all, instead of their *behavior*. Even complimenting another person with "to be" can have harmful effects. When you *rate* someone by saying "he is a good person," presumably someone else "is a bad person" by comparison—and how will you feel if you rate yourself that way? REBT categorizes rating as a form of labeling and overgeneralization [Hack #58].

Thus, perhaps we should scope the sentence this way instead, criticizing Fred's behavior rather than his total being:

I find Fred's constant nose picking and opera singing obnoxious. (On the other hand, I quite like his kindness toward animals.)

Then again, we can consider becoming aware of dogma and absolutism the main thrust of E-Prime, and permit the use of "to be," properly scoped:

> Fred's constant nose picking and opera singing are obnoxious to me.

By itself, such scoping ("obnoxious *to me*") might represent an improvement in most people's everyday speech and thought. You can treat E-Prime as a tool, whipping it out of your mental toolbox [Hack #75] and putting it away when done. You need not speak or write in E-Prime all the time, although some people do attempt to.

In Action

To use this hack, simply eliminate all uses of the verb "to be" from your communication for a given period of time, whether an hour, a day, or the time it takes you to write an email that you fear will turn into a flame.

People who speak E-Prime often choose to eliminate all forms of "to be" rather than just the so-called *is of identity* and *is of predication*. D. David Bourland, Jr. (inventor of E-Prime) and E.W. Kellogg III liken this to the situation faced by someone trying to quit smoking: although the health benefits of cutting down from two packs of cigarettes a day to just two cigarettes a day might amount to almost as much as quitting entirely, almost no one can do it. Similarly, many people who choose to speak E-Prime eliminate "to be" cold turkey because they find it easier than analyzing each use of "to be" to see whether it involves identity or predication.[2]

As a further exercise, you can try to catch writers in the use of particularly unjustified uses of "to be"; it will alert you to other people's propagandistic appeals and hone your awareness of your own use of the Verb That Dare Not Speak Its Name.

Kellogg suggests that once you get better at speaking in E-Prime, you offer your spouse or a close friend $1 every time they catch you using "to be" without immediately correcting yourself.[3] As a somewhat less costly alternative, try snapping yourself to attention [Hack #74] with a rubber band on your wrist.

You might also find that eliminating "to be" from your speech and writing forces you to express yourself more creatively. Try it! If this phenomenon interests you, you can experiment more with it by constraining yourself further [Hack #24].

In Real Life

I wrote this hack entirely in E-Prime, except for the examples. Of course, I intended this as a purposeful, rather than arbitrary constraint [Hack #24].

End Notes

1. Kellogg, E.W., III. 1987. "Speaking in E-Prime: An Experimental Approach for Integrating General Semantics into Daily Life," Etc., Vol. 44, No. 2. Reprinted in *To Be or Not to Be: An E-Prime Anthology*, published by the International Society for General Semantics, 1991. *http://learn-gs.org/library/etc/44-2-kellogg.pdf*.

2. Kellogg, E.W., III, and D. David Bourland, Jr. 1990. "Working with E-Prime: Some Practical Notes," in *To Be or Not To Be: An E-Prime Anthology*.

3. Kellogg, E.W., III. 1987.

See Also

- Wallace, Michal. The e-Primer. *http://www.manifestation.com/neurotoys/eprime.pl*. A CGI script that checks for uses of the verb "to be."

- Wikipedia entry. "E-Prime." *http://en.wikipedia.org/wiki/E-prime*. Unusually comprehensive, perhaps a result of Wikipedia's neutral point of view [Hack #64].

Learn Morse Code Like an Efficiency Expert
#53 Learn the Morse code alphabet and numbers painlessly and efficiently in an hour or less—starting now!

Frank Gilbreth, the industrial psychologist who pioneered efficiency and time-motion studies in the early 20th century, invented a quick and simple way to learn Morse code. Only the first four letters of his Morse alphabet have survived, but this hack reconstructs the rest.

In their book, *Cheaper by the Dozen*, two of Gilbreth's children, now grown, describe how they learned Morse code:

> For the next three days Dad was busy with his paint brush, writing code over the whitewash in every room... On the ceiling in the dormitory bedrooms, he wrote the alphabet together with key words, whose accents were a reminder of the code for the various letters... When you lay on your back, dozing, the words kept going through your head, and you'd find yourself saying, "DAN-ger-ous, dash-dot-dot, DAN-ger-ous."[1]

This might not be the *best* way to learn Morse code—and understanding it when it is sent to you will certainly take practice—but it is the quickest and simplest method I know of. My family members picked up roughly half the alphabet just by hearing me describe this hack while I was writing it.

In Action

Briefly, the mnemonic for each Morse code letter is a word or phrase beginning with that letter. In Table 6-1, unaccented (unstressed) syllables represent dots in Morse code, and accented (stressed) syllables represent dashes.

Letters. To use Table 6-1, first learn the alphabetic mnemonic associated
with each letter, and then reproduce the Morse code for that letter by
sounding out the stress in the mnemonic.

Table 6-1. Mnemonics for letters in Morse code

Letter	Morse code	Mnemonic	Notes
A	. -	a-BOUT	
B	- . . .	BOIS-ter-ous-ly	
C	- . - .	CARE-less CHILD-ren	
D	- . .	DAN-ger-ous	Gilbreth's list ends here.
E	.	eh?	
F	. . - .	fe-ne-STRA-tion	
G	- - .	GOOD GRA-vy!	
H	. . . .	hee hee hee hee	
I	. .	aye aye	Cheating a little, but a good mnemonic.
J	. - - -	ju-LY'S JANE JONES!	Exclamation upon learning that the famous Jane Jones will be a center-fold. Substitute "JOE JONES" if you prefer.
K	- . -	KET-tle KORN	
L	. - . .	li-NO-le-um	See *A Dictionary of Mnemonics*.[2]
M	- -	MORE MILK!	
N	- .	NA-vy	
O	- - -	OH! MY! GOD!	Said in your best Valley accent.
P	. - - .	pa-RADE PAN-el	The people who review the parade.
Q	- - . -	QUEEN's WED-ding DAY	The rhythm is the same as the opening of the familiar Wedding March, also known as "Here Comes the Bride."
R	. - .	ro-TA-tion	
S	. . .	si si si	Casual assent from a Spanish speaker.
T	-	THRUST	
U	. . -	un-der WHERE?!	Exclamation of surprise upon learning where underwear is worn.
V	. . . -	va-va-va-VOOM!	Remark upon seeing the July pictorial.

Table 6-1. Mnemonics for letters in Morse code (continued)

Letter	Morse code	Mnemonic	Notes
W	.--	with WHITE WHALE	How Captain Ahab left this world.
X	-..-	EX-tra ex-PENSE	See *A Dictionary of Mnemonics*.[2]
Y	-.--	YEL-low YO-YO	
Z	--..	ZINC ZOO-keep-er	

For example, the mnemonic for R is *ro-TA-tion*. This is the English word *rotation* with the stress on the second syllable indicated by capital letters. Because the stresses in this word run unstressed-stressed-unstressed, you know that the Morse for R is .-. (dot-dash-dot).

Numbers. Numbers must be learned somewhat differently, but because they are extremely regular, they are also relatively easy. All numbers in Morse code are five symbols (dots and/or dashes) long, and the number of dots corresponds to the numeral that is being transmitted, as shown in Table 6-2.

Table 6-2. Mnemonics for numbers in Morse code

Number	Morse code	Mnemonic notes
0	-----	0 dots
1	.----	1 dot on the left
2	..---	2 dots on the left
3	...--	3 dots on the left
4	....-	4 dots on the left
5	.....	5 dots
6	-....	(10–6) dots on the right
7	--...	(10–7) dots on the right
8	---..	(10–8) dots on the right
9	----.	(10–9) dots on the right

How It Works

This hack is yet another example of the power of mnemonics (your dear, dear friend). Remembering the dry and abstract dots and dashes of Morse code is like having to remember an arbitrary and cryptic series of commands to a computer's command-line interface. Creating English mnemonics, however, is like adding a graphical user interface on top of the command line. Simply put, it's like the difference between Windows and DOS.

In Real Life

Why would you want to learn Morse code, anyway? After all, haven't telephones and email made Morse obsolete? Not at all. Morse code is still useful in a variety of situations ranging from technological breakdown during emergencies to interfacing with the latest technology, such as texting devices.

Emergencies. You can flash Morse code with a mirror, tap it with two kinds of rock, or use many other methods to send a message far, with few resources.

Assistive technology. Anyone with minimal motor control can send Morse by using whatever they can manage—by tapping a finger or blinking, for example.

Secret communication. You can communicate via hand signals where there might be an audio bug, or by quietly tapping in a room where spoken conversation might be noticed.

Rapid communication. In 2005, the Powerhouse Museum of Sydney, Australia, held a contest between two elderly telegraph operators using Morse code and two teenagers using text messaging on their mobile phones. The telegraph operators beat them handily, despite their not using any texting abbreviations—and being 93 years old.[3] *The Tonight Show* duplicated the stunt on American television.[4]

After these contests occurred, one clever hacker wrote a free (open source) application for Nokia phones that accepts input in Morse code but sends ordinary text messages, thus allowing users to take advantage of Morse speed, even if the recipient of the message cannot understand Morse.[5]

End Notes

1. Gilbreth, Frank B., Jr., and Ernestine Gilbreth Carey. 2002. *Cheaper by the Dozen*. HarperCollins Publishers.
2. Anonymous. 1972. *A Dictionary of Mnemonics*. Eyre Methuen. The mnemonics for L and X came from this book, which contributor James Crook made me aware of after I solicited replacements for my own, fairly unimpressive L and X mnemonics. As it's a rare British book, I haven't seen it yet myself.
3. Dybwad, Barb. 2005. "Morse code trumps SMS in head-to-head speed texting combat." *http://www.engadget.com/entry/1234000463042528*.

4. Video clip from *The Tonight Show*; *http://www.makezine.com/blog/archive/2005/05/video_morse_cod.html*.

5. "Morse Texter." 2005. *http://laivakoira.typepad.com/blog/2005/05/morse_texter.html*.

See Also

- See the excellent Morse code page in the Wikipedia for information on niceties such as word and sentence spacing and timing, punctuation, special symbols, accented letters, abbreviations, and so on: *http://en.wikipedia.org/wiki/Morse_code*.

HACK #54 Harness Stage Fright

At some point, nearly everyone has to speak to a group about something, but most of us find ourselves overwhelmed with fear when the time comes. However, if you reduce the fear to a manageable level, you can channel its energy into making your presentations more powerful.

If someone walked up to you today and asked you to give a lunchtime talk about your favorite hobby tomorrow, how would you feel? If you're like most people, you'd probably think fast about an excuse to get out of it, and if you couldn't, you'd lose sleep tonight. Public speaking is terrifying to many people, even in such a low-pressure setting and with a topic that we find pleasant.

Because so many people fear public speaking, overcoming and harnessing that fear can give you a distinct advantage. It's a powerful skill, and it comes into play for almost everyone at some time in their lives. Most businesspeople will be called on to give a presentation about something in their careers, for example. Even if you don't have that kind of job and aren't an actor, stage fright can sneak up on you if you attend a meeting of your church or neighborhood association, take a class and need to ask questions, or decide to lead a Girl Scout troop.

Even writer's block can be a form of stage fright, proving that you don't need to face a roomful of people to use these techniques.

In Action

The best basic strategy to handling stage fright is to first reduce it to a manageable level and then to use what's left to make your performance more powerful—to give you "stage presence." You should perform some of the following techniques well ahead of time, some soon before you speak, and some right before you start and during your performance.

Make notes. As far ahead of time as you can, make notes and know them. Take time to organize your thoughts ahead of time. You'll reap huge benefits from this in terms of reducing your fear when it's time to speak. You don't have to write the whole thing out (and sometimes you won't really have time), but even a few keywords or an outline will help you find your place if you start to panic. It will also help you be sure that you say everything you want to say without repeating yourself and that you present your ideas in a logical order.

Don't procrastinate about making your notes, because you'll want to give yourself some time to use them in preparation. After you make your notes, take some time to go over them so that you know what's there.

Pay special attention to the beginning and end of your talk and any transitions between sections. Knowing the beginning well will help get you over the obstacle of getting started without panicking; transitions will help you move on if you need to; and knowing the end will help you leave the audience with a good impression, no matter what happens in the middle of the presentation.

Imagine pitfalls. Think about things that could possibly go wrong and figure out an emergency plan. Some people avoid thinking about anything that could go wrong because they think it will make them panic. It can be a little scary to imagine potential problems, but there are two main reasons to do so. The first is to use it as a "dress rehearsal" so that you can figure out ahead of time what you might do in case of trouble. Then, you don't have to worry so much about the problems, because if one crops up, you already know how to react.

The second reason to imagine what could go wrong is to gain perspective. When it comes down to it, the worst that can happen is usually not that bad. Seldom is a life hanging in the balance on the effects of what you're getting ready to say. Most of the time, the worst that can happen is that you'll be slightly embarrassed if something goes wrong. Given how many times all of us are embarrassed in our life, that's hardly a dire threat. Some of the other hacks in this book can be tremendously helpful in disarming irrational fears, such as learning about the ABC model of emotion [Hack #57] and learning to avoid cognitive distortions [Hack #58].

Remove sources of stress. Anticipate trivial and tangential sources of stress and anxiety, and remove them in advance. In other words, don't make it any harder than it has to be. Make sure you get enough sleep the night before you speak. Make sure you eat a little beforehand, but not too much. Make sure your clothes are clean and appropriate, and that your appearance is acceptable.

Take care of all the little details that can leach away your attention and energy, so you can focus on the task at hand. Many of these basic concerns, such as paying attention to sleep, nutrition, and exercise [Hack #69], will also improve your general brainpower.

Consider your audience realistically. Remember that the audience wants to be on your side—really! We imagine that when we speak, the audience is ready to criticize and humiliate us, waiting for us to make a mistake, poised to laugh at us. In fact, most of the time, people relate to you when they watch you and hope you'll do well.

To find evidence, you don't have to go any further than your own imagination. When you see someone speak or act, and they blunder or forget their lines, don't you find yourself holding your breath for them, and aren't you happy and relieved when they recover and go on to do well? Most people are like you: they are basically decent people who don't want to see anyone hurt or humiliated. They'll feel the same when you're the one in the spotlight.

Keep perspective. Take yourself seriously, but not too seriously. If you're going to speak, you have a reason to do so that matters, either to you or to someone else. Respect yourself or whoever asked you to speak, and approach your task with some appreciation that you'll make an impression. Don't be diffident and don't apologize for speaking at all, or for minor glitches such as coughing or stumbling on a word; doing so only takes up time and puts the focus on the mistake, not on what you're there to do.

On the other hand, again, most of the time your speech isn't a life-or-death matter, and it's certainly not going to change history if your tongue trips. Be ready to brush such things off and see the humor in your situation, and you'll keep the audience on your side.

Mind your body. Just before you go on, work out physical kinks a little. Don't go into your speech out of breath, but take a moment to stretch, breathe deeply, and even take a little walk before you start. You'll relax your body, improve your blood flow, increase oxygen to your brain, and make yourself more alert.

Feel your fear. With preparation, you've reduced your fear to a more manageable level. Now, go ahead and let yourself feel it a little. Let yourself get excited and feel it physically. Try not to panic, but go ahead and let through some of the sensations.

Reframe your fear. Rethink your fear-induced physical sensations. This is possibly the trickiest part of the hack, and the heart of it.

Now that you can feel some arousal, try to turn it into excitement. Pull your thoughts away from your fears and fix your imagination on your positive goals. Do you want to change someone's mind with your arguments? Do you want to express something that's important to you? Do you want to teach people about something you love? Remember that you have an opportunity now, and isn't that great? Remember what's causing you to go through with this, even though you're scared; there must be something behind it that you care about quite a lot.

Slowly, as you realize and picture everything good that could happen as a result of your performance, you'll feel the fear changing. The feelings won't go away completely, but with luck and focus, they will turn into a happier sort of excitement, the kind you feel before you do something fun. Blend your passion for your subject matter and your hoped-for results with the excited feelings, and let that fill you and carry you forward. Again, the ABC model of emotion [Hack #57] can help you do this with more sophistication and control.

Use good body language. Be aware of what you're projecting with your physical self. Hold your head up, walk with confidence, smile if it kills you, and don't fidget. Looking the part helps you to project your energy as positive confidence and command of the stage, and that's what your audience will see, even if you're still scared inside.

Fake it till you make it. In other words, *act* like a confident speaker, no matter how you feel. Pretend you're anyone you admire, move as you imagine a confident person moves, speak like that person would speak, and so on.

Not only will you fool a lot of people into believing that you know exactly what you're doing, but also the effect will work on you, too. Stepping into that person's shoes actually makes you feel more confident and able as well.

Keep breathing. Force yourself to pause between sentences to take a good deep breath from time to time while you're speaking. You might think this takes a long time and looks silly, but the odds are that you're speaking too fast anyway.

Pausing for breath forces you to slow down and think about what you're doing, check your notes, and generally be present in the moment. Many times, your audience will read your breath as a dramatic pause anyway, which gives more power to what you're saying. At the very least, it gives them a moment to take in and understand your words. It can also help you reduce stuttering and bad speaking habits like filling every possible pause with "um" or some other sound.

Move on. If something goes wrong, pick up and go on. Dwelling on a mistake only magnifies it. You don't want to steal focus from what you're there to talk about, so correct the mistake if necessary, and move away from it as quickly as possible. Don't fumble, lose your concentration, apologize, or panic further; just move on.

Finish well. Finish as well as you possibly can. No matter what kind of disasters you believe have happened while you were onstage, make the best impression you can at the end. Plan to end your talk with a good story or a strong argument. Never end by saying, "I guess that's about it," or something similarly mushy.

Finally, after you've said the last thing you have to say, stop talking, look into the audience, and smile. You can then take questions, walk confidently away, sit down, or do whatever you need to do. Just be sure you remain "in character" until you really are off the stage, in whatever form that takes.

How It Works

Why does stage fright grip us so tightly and irrationally? Probably because we form our feelings about it in childhood. Young children speak to people freely and express their thoughts as well as they can, without reticence. However, most of us got our first experience speaking to a group at school. We probably had in the audience a teacher ready to grade our performance and classmates who would certainly make fun of it if we made a mistake or looked silly at all—partly because making fun of us would help quiet their own fears about speaking.

So, we learned to believe that if someone in authority watched us, we would be judged, and that the rest of the audience was hostile and waiting to pounce on any slip-up. Reminding ourselves that we're reacting to old information that's no longer relevant to the present situation can help quell those deep-seated fears.

The technique of channeling stage fright into performance power stems from the idea that the body's state of arousal is interpreted by the mind within a framework of emotional reaction and knowledge about the current situation. Under stress, the body creates a surge of adrenaline, pounding heartbeat, heightened senses, faster breathing, and related reactions. Whether we feel those physical effects as fear, exhilaration, sexual arousal, rage, or something else depends on how we feel at the time and what we believe is going on. It's part of the reason that some people are turned on sexually by danger or pain, or have a good time at a scary movie or on a frightening roller coaster.

Psychologists Schachter and Singer showed that situational factors cause us to frame physiological effects.[1] They gave a mild stimulant to a group of subjects and then placed some of the subjects with a person who exhibited anger and others with someone exhibiting happiness. In both groups, more of the people who had been given the stimulant self-reported strong feelings mirroring the "emotional leader" than those who hadn't received the stimulant that produced the physiological effects of arousal.

A similar experiment by Dutton and Aron showed how the mind can shift the effects from one kind of arousal to another.[2] In this study, an attractive woman interviewed men, some on a swaying rope bridge 200 feet over a river and some on firm ground. During the course of the interview, she gave the men her phone number. More than 60% of the men who talked to her on the rope bridge phoned her afterward, compared to 30% of the men who talked to her on the ground. They had interpreted the heightened arousal level that was produced by the more dangerous situation as greater attraction.

With the techniques in this hack, you're replicating the effects of these studies intentionally. By consciously reframing physical sensations, we can often convert one emotional reaction to another, in this case transforming fear into excitement that we can transmit through our performance.

In Real Life

I learned a great deal about acting and public speaking in high school and college. I performed in community and semi-professional theater, took acting classes, and competed in public speaking tournaments. I didn't grow up to be an actor or a professional speaker, but I've found the experiences I had and techniques I learned to be invaluable in later life. To name only a few situations where the skills have served me well, I've used them to:

- Teach informal classes
- Participate well in classes as a student

- Make a good impact in job interviews
- Conquer social anxiety in meeting new people
- Persuade groups about issues I found important

In fact, these techniques have become so much a part of me that it was difficult for me to dissect them to write the hack! I consider it some of the most important training I've ever had. If you're a parent, this is also a strong endorsement to encourage your kids to study some acting, debate, and speech techniques while they're young.

End Notes

1. Schachter, S., and J. E. Singer. 1962. "Cognitive, social and physiological determinants of emotional states." *Psychological Review*, 69: 379–399.
2. Dutton, D. G., and A. P. Aron. 1974. "Some evidence for heightened sexual attraction under conditions of high anxiety." *Journal of Personality and Social Psychology*, 30: 510–517.

—*Marty Hale-Evans*

H A C K #55 Ask Stupid Questions

At school or at work, we often feel as though we are "drinking from the firehose" when we have to learn a new extensive or complicated subject or task. In these situations, the least stupid thing we can do is ask stupid questions.

Human beings have acquired several different kinds of learning during the course of our evolution. Being able to acquire short-term disposable knowledge (such as where we left our spear) without cluttering up our brain with useless trivia might have been just as important to our survival as learning the long-term skills (such as building a fire or knowing when to plant crops) commonly associated with survival. As modern humans, if we learn something and then forget it before we want to, it might be because our brain never indexed it for long-term use.

Often, this occurs because information is coming at us so fast that our brain just can't keep up. Studies have shown that our short-term memory can in fact hold only between five and nine items [Hack #11], depending on the information.[1] Thus, if your short-term memory is full and you are given new items to learn, your brain will be forced to sacrifice. This hack provides some advice that will help you control the incoming flow of information so that your brain has a chance to index it in your long-term memory.

Learning is something we *do*, not something we *receive*. When we are tasked with learning some large, new body of knowledge, it is up to us to control the flow of information we are receiving, to the best that the situation allows. This will allow our brain to process the information and build the deep structures necessary for true *comprehension*—learning that stays with us, that we can readily recall in the future.

Failure to stem the flow of input can result in frustration, stress, or short-term learning that can't be called up the next time we need to apply it. It wastes not only our own time, but also the time of the person teaching us.

In Action

In his book *Surely You're Joking, Mr. Feynman!*, Nobel prize-winning physicist Richard Feynman writes about his experience teaching a group of students in Brazil:[2]

> One other thing I could never get them to do was to ask questions. Finally, a student explained it to me: "If I ask you a question during a lecture, afterwards everybody will be telling me, 'What are you wasting our time for in the class? We're trying to *learn* something. And you're stopping him by asking a question.'"

Because all the students were afraid to ask questions, Feynman got highly frustrated and came away convinced that the students were not actually learning anything at all. Too often, in a learning situation, we tend to just nod and allow the instruction to progress without adequate comprehension. Even when we are in a one-on-one learning situation with just an instructor, the temptation to sit passively often results in a monologue. We do this for several common reasons, each of which is a fallacy:

Excuse	Reality
We don't want to look stupid or waste the class's time by asking for something to be repeated.	We will look even more stupid if the instructor sees that her time was wasted.
There's a lot of material to cover, and we don't have time to waste.	We are wasting more time, perhaps even the whole session, if what we're learning fails to stay with us.
We *kind of* think we understand a concept and assume that it will "gel" later.	Too often, this momentary learning falls by the wayside as we try to take on new knowledge.
The instructor might hit upon the part that we don't understand at a later point.	What if she doesn't?

So, what should you do? A much better method is to be active, not passive:

- Interrupt your instructor at every step if you are at all uncertain about what you are learning.

- Ask stupid questions, even if you feel it's something you should have picked up already and it might make you look bad.

- Make your instructor start again—from the beginning, if necessary—if your understanding is not complete.

- Don't leave *dangling pointers*—points that you didn't quite understand, but for which you are anticipating further elucidation. Dangling pointers have a tendency to get lost.

How It Works

Whether we are studying quantum chromodynamics or just making a new friend, a similar process is occurring in our brain. We are building a mental model of the subject we are learning so that we can run mental simulations against it to predict future behaviors. Indeed, the model we build internally should be robust enough for us to predict behaviors that were never described during the initial learning process, if it is to be considered anything other than rote learning.[3]

Since there is no way to import mental models directly into the brain (not yet, at least!), we must build them piecemeal, as we might assemble a model airplane. To continue the analogy, the instructor is handing you pieces of the model, and you are assembling them internally. If the instructor is handing you pieces too quickly, you will not have time to assemble them. Further, if you assemble part of the model but are unsure about it, there is no point in continuing to add pieces.

In Real Life

At your new job, you are tasked with helping to maintain a large, complex web site. There is a lot to learn: database structures, passwords, server locations, error-reporting procedures, source code revision-control systems, build procedures, and so on. There's some documentation, but as is usually the case, it hasn't been kept up-to-date. So, other employees who are familiar with the project have to teach you.

You are eager to impress your new colleagues, so you start learning your new project with vigor. The person walking you through it all is initially impressed as you quickly fly through the different areas you are supposed to learn. You might have missed some finer points, but you don't want them to think they've made a mistake in hiring you, so you stay mum.

You can still repeat back some of the information you've been taught, at least at first. But as the days go by, your initial comprehension seems to vanish in the wind, and your co-workers become less impressed as they have to start answering questions about things they've already told you about.

What should you have done? You will ultimately be judged by how quickly you are able to contribute to the project, not by how knowledgeable you seem up front. Stupid questions are to be expected, and would have quickly been forgotten. When you've had enough for one day, call it quits, and go back and review what you've learned so far. Continuing is just a waste of your and your co-workers' time, and will ultimately reflect poorly on you.

End Notes

1. Miller, George A. 1956. "The Magical Number Seven, Plus or Minus Two." *The Psychological Review*, 63: 81–97.

2. Feynman, Richard P. 1985. *Surely You're Joking, Mr. Feynman!* (page 194). Bantam Books.

3. Minsky, Marvin. 1985. *The Society of Mind.* Simon & Schuster, Inc. A collection of short, readable essays detailing how the human brain builds up mental models and how they might be simulated on a computer.

—*Mark Schnitzius*

HACK
#56

Stop Memory-Buffer Overrun

The length of a sentence isn't what makes it hard to understand; it's how long you have to wait for a phrase to be completed.

When you're reading a sentence, you don't understand it word by word, but rather phrase by phrase. Phrases are groups of words that can be bundled together, and they're related by the rules of grammar. A noun phrase will include nouns and adjectives, and a verb phrase will include a verb and a noun, for example. These phrases are the building blocks of language, and we naturally chunk sentences into phrase blocks just as we chunk visual images into objects.

This means that we don't treat every word individually as we hear it; we treat words as parts of phrases and have a buffer (a very short-term memory) that stores the words as they come in, until they can be allocated to a phrase. Sentences become cumbersome not if they're long, but if they overrun the buffer required to parse them, and that depends on how long the individual phrases are.

In Action

Read the following sentence to yourself:

- While Bob ate an apple was in the basket.

Did you have to read it a couple of times to get the meaning? It's grammatically correct, but the comma has been left out to emphasize the problem with the sentence.

As you read about Bob, you add the words to an internal buffer to make up a phrase. On first reading, it looks as if the whole first half of the sentence is going to be your first self-contained phrase (in the case of the first, that's "While Bob ate an apple")—but you're being led down the garden path. The sentence is constructed to dupe you. After that phrase, you mentally add a comma and read the rest of the sentence...only to find out it makes no sense. Then you have to think about where the phrase boundary falls (Aha, the comma is after "ate," not "apple"!) and read the sentence again to reparse it. Note that you have to read it again to break it into different phrases; you can't just juggle the words around in your head.

Now try reading these sentences, which all have the same meaning and increase in complexity:

- The cat caught the spider that caught the fly the old lady swallowed.
- The fly swallowed by the old lady was caught by the spider caught by the cat.
- The fly the spider the cat caught caught was swallowed by the old lady.

The first two sentences are hard to understand, but make some kind of sense. The last sentence is merely rearranged but makes no natural sense at all. (This is assuming it makes some sort of sense for an old lady to be swallowing cats in the first place, which is patently absurd, but it turns out she swallowed a goat, too, not to mention a horse, so we'll let the cat pass without additional comment.)

How It Works

Human languages have the special property of being recombinant. This means a sentence isn't woven like a scarf, where if you want to add more detail you have to add it at the end. Sentences are more like Legos. You can break them up and combine them with other sentences, or pop them open in the middle and add more bricks.

Have a look at these rather unimaginative examples:

- This sentence is an example.
- This boring sentence is a simple example.
- This long, boring sentence is a simple example of sentence structure.

Humans understand sentences by parsing them into phrases. One type of phrase is a noun phrase, the subject or object of the sentence. In "This sentence is an example," the noun phrase is "This sentence." For the second, it's "This boring sentence."

Once a noun phrase is fully assembled, it can be packaged up and properly understood by the rest of the brain. During the time you're reading the sentence, however, the words sit in your verbal working memory—a kind of short-term buffer—until the phrase is finished.

> There's an analogy here with visual processing. It's easier to understand the world in chunks—hence the Gestalt Grouping Principles. With language, which arrives serially, rather than in parallel, like vision, you can't be sure what the chunks are until the end of the phrase, so you have to hold it unchunked in working memory until you know where the phrase ends.

Verb phrases work the same way. When your brain sees "is," it knows a verb phrase is starting and it holds the subsequent words in memory until that phrase has been closed off (with the word "example," in the first sentence in the previous list). Similarly, the last part of the final sentence, "of sentence structure," is a prepositional phrase, so it's also self-contained. Phrase boundaries make sentences much easier to understand. The object of the third example sentence, instead of being three times more complex than the first (it's three words, "long, boring sentence," versus one, "sentence"), can be understood as the same object, but with modifiers.

It's easier to see this if you look at the tree diagrams shown in Figure 6-1. A sentence takes on a treelike structure, for these simple examples, in which phrases are smaller trees within that. To understand a whole phrase, its individual tree has to join up. These sentences are easy to understand because they're composed of very small trees that are completed quickly.

We don't use just grammatical rules to break sentences in chunks. One of the reasons the sentence about Bob was hard to understand was that you expect, after seeing "Bob ate," to learn about *what* Bob ate. When you read "the apple," it's exactly what you expect to see, so you're happy to assume it's part of the same phrase. To find phrase boundaries, we check individual

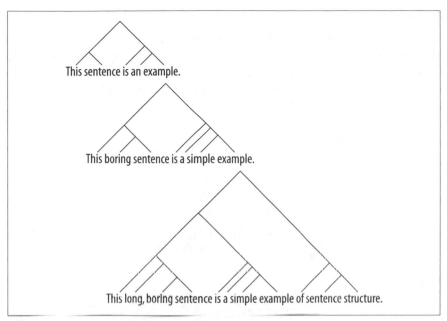

Figure 6-1. How the example sentences form trees of phrases

word meaning and likelihood of word order, continually revise the meaning of the sentence, and so on, all while the buffer is growing. But holding words in memory until phrases complete has its own problems, even apart from sentences that deliberately confuse you, which is where the old lady comes in.

Both of the first remarks on the old lady's culinary habits require only one phrase to be held in buffer at a time. Think about what phrases are left incomplete at any given word. It's clear what any given "caught" or "by" words refer to: it's always the next word. For instance, your brain read "The cat" (in the first sentence) and immediately said, "did what?" Fortunately the answer is the very next phrase: "caught the spider." "OK," says your brain, and it pops that phrase out of working memory and gets on with figuring out the rest of the sentence.

The last example about the old lady is completely different. By the time your brain gets to the words "the cat," three questions are left hanging. What about the cat? What about the spider? What about the fly? Those questions are answered in quick succession: the fly the old lady swallowed; the spider that caught the fly; and so on.

But because all of these questions are of the same type, the same kind of phrase, they clash in verbal working memory, and that's the limit on sentence comprehension.

In Real Life

A characteristic of good speeches (or anything passed down in an oral tradition) is that they minimize the amount of working memory, or buffer, required to understand them. This doesn't matter so much for written text, in which you can go back and read the sentence again to figure it out; but you have only one chance to hear and comprehend the spoken word, so you'd better get it right the first time around. That's why speeches written down always look so simple to understand.

That doesn't mean you can ignore the buffer size for written language. If you want to make what you say, and what you write, easier to understand, consider the order in which you are giving information in a sentence. See if you can group together the elements that go together so as to reduce demand on the reader's concentration. More people will get to the end of your prose with the energy to think about what you've said or do what you've asked.

See Also

- Caplan, D., and G. Waters. 1998. "Verbal Working Memory and Sentence Comprehension." *http://cogprints.ecs.soton.ac.uk/archive/00000623*.
- Steven Pinker discusses parse trees and working memory extensively in *The Language Instinct: The New Science of Language and Mind* (Penguin Books Ltd.).

—*Matt Webb and Tom Stafford*

CHAPTER SEVEN

Clarity
Hacks 57–65

Unlike most of the other chapter titles in this book, this one might not be self-explanatory. What is clarity? Clarity is freedom from unpleasant thoughts, emotions, and states of consciousness. Really, it is freedom pure and simple, but for the purposes of this book, it is the freedom to *think*, and developing that is a crucial branch of the mental arts.

You can't think well if you are angry, depressed, or frightened, or if your mind is cluttered with thoughts. Most computers' hard drives need to be defragmented periodically. Your mind is no different in this respect.

This chapter explores various ways to defragment your mental hard drive and clear its desktop. At first glance, you might think that some of these hacks are mystical nonsense, but there's no incense burning or bell ringing in this chapter—only stuff that works.

HACK #57 Learn Your Emotional ABCs

Mental and emotional clarity reinforce each other, so don't ignore your emotions in your quest to be a better thinker. Greater clarity is just a few steps away.

The *ABC* model of emotion, widespread in contemporary psychotherapy, holds that it is not an *activating* (A) event, such as rejection by a friend or lover, that causes you emotional *consequences* (C) such as depression; rather, the linchpin is your invisible *beliefs* (B) about the event that come in between A and C. Fortunately, it's often easier to intentionally change beliefs than emotions.

Since at least the time of the ancient Stoics, some have believed that our *circumstances* don't control whether we're happy, but our *thoughts* about them do. Our reasoned thoughts and beliefs form a kind of buffer between reality and our private selves—in theory. In practice, our thoughts often don't *buffer* us from events we don't like so much as *amplify* those experiences, causing us emotional turmoil and suffering. In fact, our thoughts can be so irrational and so removed from reality that they often make us suffer, even when nothing is objectively wrong.

Questioning the irrational thoughts that cause you emotional pain and thereby cloud your reasoning can help you think more clearly and act more effectively. This hack explores the ABC model of emotion pioneered by Albert Ellis, developer of Rational Emotive Behavior Therapy (REBT). You can use it as a lever to heave off the massive boulders of emotional self-oppression.[1]

Going Beyond the ABCs

According to Ellis, people in our culture go through three *normal* stages of emotion (shown in Table 7-1) many times a day, consisting of an activating event in the external world, filtered through their beliefs, resulting in emotional consequences.

Table 7-1. The ABCs of emotion

Stage	Name of stage	Description
A	Activation	The triggering event
B	Beliefs	What you told yourself about A
C	Consequences	The emotional results of B: how you reacted to your belief

Engaging in the three additional steps of *rational self-analysis*[2] shown in Table 7-2 can help you be more reasonable, and even happier.

Table 7-2. Three additional stages

Stage	Name of stage	Description
D	Disputation	The arguments you will make to yourself, and new beliefs you will form, to help you achieve E
E	Effect	The new effect you want: how you'd *prefer* to feel and behave
F	Further action/effective new philosophy	Further steps you will take to avoid the same dysfunctional thoughts and reactions in the future

The steps of rational self-analysis also have an alphabetic mnemonic. They constitute the three stages beyond ABC, if you take them: disputation of your irrational beliefs, aiming at an emotional effect, followed by further action to stop the cycle from occurring again.

Disputing Irrational Beliefs

You can use three broad classes of disputation in a rational self-analysis:

Empirical
> What evidence is there for this belief? Is there a law of nature that proves it, or does the *law* exist only in my mind?

Logical
> Just because I want something, does it follow logically that I *must* get it? Just because it's uncomfortable, does it follow that it's *awful*? Just because I made a mistake, does it follow that I'm an *idiot*?

Pragmatic
> Does believing this help me to be effective and happy? Or does it create interpersonal problems, roadblocks, and stress?

You can also dispute your beliefs with a list of cognitive distortions [Hack #58]. Sometimes just being able to nail one of your irrational beliefs as a classic example of one of the distortions on the list is enough to make it shrivel in a smoking heap.

In Real Life

In practice, you will usually perform a rational self-analysis in the order C, A, B, E, D, F. That is, you'll do the following:

1. Experience some emotional *consequences* (C).
2. Identify the *activating event* (A).
3. Identify which *beliefs* (B) the event was filtered through.
4. Determine what *effect* (E) you *would* have preferred.
5. To that end, *dispute* (D) your old beliefs and create some new ones.
6. Make plans for *further action* (F).

Let's follow the steps of the typical rational self-analysis of a hypothetical grad student. We'll call this person Russell, after Bertrand Russell, the famous rationalist. Our man Russell has a disability that sometimes causes him some discomfort and embarrassment, and he is trying to learn to cope with it.

One night at dinner, Russell has a sudden attack of an incurable disease called *hodaddia*, which he contracted from a mosquito bite on a tropical vacation several years ago. When the attack struck tonight, he collapsed in pain and turned bright mauve all over, as usual. This is the *activating event*.

Russell feels furiously ashamed and embarrassed. He loathes and pities himself, and then he gets depressed and needy, feeling helpless. These are the *emotional consequences*. He lies in bed ineffectually and demands that his roommates brew him big pitchers of hot lemonade with maple syrup, which is the only thing besides his medication that seems to bring relief.

Finally, Russell decides he's had enough. Just this once, he's going to analyze his thinking and clear his head. He identifies A and C.

Next, Russell dredges up his *beliefs* (B) that transformed the activating event (A) into the emotional consequences (C):

- I'm weak and worthless.
- I'm crippled by this hodaddia; as a hodaddia victim, I can never do anything worthwhile, because hodaddia can strike at any moment.
- Everyone hates me, because they think I'm a purple freak.
- I probably ruined any chance of seeing my date again when I turned mauve and started moaning at dinner.
- I should never have gone sunbathing in my Speedo on spring break instead of staying in my mosquito-proof tent. I'm such an idiot.

The *effects* (E) Russell wants to have in the future, instead of his normal emotional consequences, are as follows:

- Feel calm and reassured after an attack
- Get on with his life
- Stop bugging his roommates to take care of him
- Be more effective again

To achieve these effects, Russell *disputes* (D) his irrational beliefs as follows:

1. How does it *logically* follow that just because I have a disability, I'm "weak and worthless"? No, that's just *labeling* myself. In fact, lots of people with disabilities throughout history have been incredibly strong and creative, and produced work of great worth; consider Stephen Hawking and Vincent Van Gogh.

2. Believing I can never do anything worthwhile is *binary thinking*. Even if my hodaddia were much worse, I could still do *some* worthwhile things. *Empirically*, I've already managed to do some interesting things, such as my blog, which has thousands of subscribers, and the Obfuscated INTERCAL programming contest, which I've won for the last four years. Plus, I'm a good friend, when I'm not demanding hot lemonade with maple syrup. Why should I *filter out* all the good stuff I do when I think about my life?

3. Granted, a lot of people think I'm weird because I turn purple and moan sometimes, but not everyone knows about my hodaddia disability or would hold it against me if they did. I am blessed with friends and family who love and understand me to various degrees, but even if no one did, I could still find pleasure in life and be happy, so my whole mental debate about whether people hate me because I'm a "purple freak" can just sod off!

4. As for my date tonight: how long could I keep my hodaddia problem from Chris? It's better to be open with people I'm going to be romantically involved with; if I scare them off, that's one less shallow person to have in my life. Anyway, I don't have *telepathy*; if I want to know what Chris really thought and felt, I should call and find out, instead of assuming the worst. Maybe Chris is worried about me and would like to know I'm OK.

5. Should, should, should. This is just *musturbation*. Even if I was a fool to go sunbathing, what law of the universe says that no one is allowed to be a fool? If there is one, a lot of people are breaking it. Anyway, none of my friends who went out sunbathing that spring break caught the hodaddia virus. I just happened to have a hidden genetic susceptibility. How could I know that? So, unless I want to be miserable for the rest of my life, I *should* stop engaging in this *shouldy* thinking.

6. Anyway, *pragmatically* speaking, all of this self-loathing, depression, and being demanding is not at all helpful. It drags me down and makes me ineffective and ineffectual, which is itself one of the things I'm berating myself about. So, I had better uproot all these irrational beliefs.

Finally, Russell decides to take some *further action* (F):

- Write up his rational self-analysis in a more portable form, such as an index card he can carry in his exoself [Hack #17] and refer to the next time he starts hating himself

- Continue to extend and deepen this self-analysis so that he understands himself and his problems better

- Join some hodaddia support groups online and read up on further action he can take, such as educating friends and acquaintances about his condition, and then take those actions

Suddenly, Russell doesn't feel so bad anymore. In fact, he feels like working on a new project.

End Notes

1. Ellis, Albert, Ph.D., and Robert A. Harper, Ph.D. 1997. *A Guide to Rational Living*, Third Edition. Wilshire Book Company. This is the classic introduction to Rational Emotive Behavior Therapy. The second edition, titled *A New Guide to Rational Living*, was written in E-Prime [Hack #52].

2. Froggatt, Wayne. 1997. *GoodStress: The Life That Can Be Yours*. HarperCollins Publishers (New Zealand) Limited.

See Also

- Froggatt, Wayne. 1997. "Twelve Rational Principles: Using the principles of Rational Effectiveness Training to achieve a satisfying and productive life." *http://www.managingstress.com/articles/frogatt.htm*. This page contains the first chapter of *GoodStress* and is a good introduction to Rational Emotive Behavior Therapy.

HACK #58   Avoid Cognitive Distortions

Learn to avoid 15 mental mistakes that distort your emotions and in turn further distort your thinking.

This hack will help you recognize the kinds of irrational thinking that you, like everyone else, tend to engage in from time to time. Reducing your irrational thinking will often help you to become a happier, and therefore more *rational* and *effective*, person.

The techniques in this hack can stand alone—after you recognize these thoughts for what they are, they lose much of their power—but the techniques work better when you use them as part of the ABC model of emotion [Hack #57], so I recommend that you read that hack first. These distorted thoughts tend to occur at the B stage of that ABC model.

Three Core Irrational Beliefs

The primary cognitive distortions of *self*, *others*, and *world* are the foundations for many of the secondary cognitive distortions in the next section. You might think of them as *clumps* of thought—that is, distorted viewpoints, orientations, or ways of approaching your life rather than individual thoughts.

Table 7-3 presents some sample distortions and common triggers for each of these perspectives.[1]

Table 7-3. Irrational beliefs and their common triggers

Orientation	Example	Common triggers
Self	"I must be absolutely perfect in everything I do; otherwise, I can't stand myself because I am completely worthless!"	Failing a test, missing a deadline, creating an imperfect work of art (i.e., *any* work of art)
Others	"Everyone I meet must treat me just the way I like; otherwise, they are completely worthless!"	Getting stuck in line, receiving bad service in a restaurant, being spurned in love
World	"World events must happen just the way I want them to; otherwise, *life sucks* and is completely unbearable!"	Wars, elections, famines, epidemics, Trials of the Century, buggy software, the weather, insufficient parking, most other world events that might not go the way you prefer

Most people fall prey to *all* of these irrational beliefs from time to time, and most people have a "favorite" one. For example, the viewpoint that has most often ruined my day is the World example.

12 Misinterpretations

Whereas you might imagine the irrational beliefs in the previous section as big musical themes that run throughout people's lives, the smaller irrational beliefs shown in Table 7-4 are more like the musical notes that make up those themes.[2,3]

Table 7-4. Examples of misinterpretation

Misinterpretation	Description	Example
Awfulizing	Judging something uncomfortable or unpleasant to be *awful* or *terrible* (see *minimization*).	"My hard drive got corrupted. I lost two hours of work. This is *totally horrible! Totally!*"
Binary thinking	Thinking in absolutistic, black-or-white, all-or-nothing terms.	"There are only two kinds of people in the world: SF fans and mundanes. Don't even bother with the mundanes."
Disqualifying	Insisting that positive experiences "don't count."	"Yeah, I once graduated *cum laude*, but what does that matter *now*?"
Emotional reasoning	Believing that your emotions are telling you the truth about reality.	"Uh oh, I have a horrible sinking feeling. What am I overlooking about this contract?"

Table 7-4. Examples of misinterpretation (continued)

Misinterpretation	Description	Example
Filtering	Seeing only the negatives in a situation while ignoring the positives.	"My marriage is falling apart. When was the last time she told me she loved me? I'm not *talking* about the poems and mix CDs she made me or the fact that she fixed my computer, or took care of my kid…" (This is related to *disqualifying*.)
Labeling	Judging yourself or another person to be a *jerk, idiot, scumbag*, etc., because you don't like something the person did (a form of *overgeneralization*).	"That utter *bullet head* in the monster pickup who cut me off probably has testosterone Jell-O for brains! I hope he has a wreck!"
Minimization	Judging something positive to be vanishingly unimportant (see *awfulizing*).	"My blog is really popular, but it's hardly my life's work. I need to stop wasting my time with this crap."
"Must" statements	Applying words such as *should, must,* and *ought* to yourself or others. When you do it to yourself, you often feel guilty; when you do it to others, you often feel anger (also known as *shoulding all over yourself*).	"I *should* have gotten an A on that paper. Maybe I'm stupid. No, the professor is a jerk. He *should* have given me that A; I deserved it."
Overgeneralization	Believing that something has always happened and always will; assuming that one event is representative of the entire situation, forever.	"I *always* hate going to parties with my sister. No one *ever* talks to *me*; what's the point?"
Personalization	Assuming without evidence that an event is connected with you.	"My boss just went into her boss's office and closed the door. They're obviously talking about firing me."
Precognition	Believing that you can predict the future.	"I just *know* I'm going to get stuck in traffic and miss the show."
Telepathy	Believing you can know what other people are thinking without asking them.	"Peter probably hates me, so why should I bother flirting with him?"

Once you recognize these beliefs in yourself, you'll find that it's a relief to stop whistling those tunes.

In Real Life

Now that you have an idea of these little mental land mines, what can you do to avoid them or, better, defuse them?

Just recognizing them helps, but you can also use rational self-analysis [Hack #57] to mentally *dispute* your distorted thinking and replace it with healthy thoughts that will further your projects and make you happier.

Yes, *happier*. Some people believe that we cause our feelings when we think rationally or irrationally. This means that you can't *make* anyone else happy—too bad. On the other hand, no one else can *make* you unhappy—only yourself. So see if you can put an end to it!

End Notes

1. Ellis, Albert, Ph.D., and Robert A. Harper, Ph.D. 1997. *A Guide to Rational Living*, Third Edition. Wilshire Book Company. The classic introduction to Rational Emotive Behavior Therapy. Buy this book if you can buy only one off this list.

2. Burns, David, M.D. 1999. *Feeling Good: The New Mood Therapy*. Avon Books. A popular classic; especially good for combating depression.

3. Bambrick, Leon. 2004. "How to Be Depressed: A Quick Guide to Getting Less Out of Life." *http://www.secretgeek.net/depression_is_easy.asp*. A satirical look at cognitive distortions from a software developer's standpoint.

See Also

- Froggatt, Wayne. 1997. *GoodStress: The Life That Can Be Yours*. HarperCollins Publishers (New Zealand) Limited. Good general book on Rational Emotive Behavior Therapy. Out of print, but the first chapter is online: *http://www.managingstress.com/articles/frogatt.htm*.

HACK #59 Use the Fourfold Breath

The Fourfold Breath is a long-known method of rhythmic breathing that helps you calm your body down so that you can think clearly.

The Fourfold Breath, a kind of *pranayama yoga*, is an effective brute-force method of calming your body by consciously controlling your rate and pattern of breathing. It's simple to learn, and easy to do and remember.

You can use the Fourfold Breath as an adjunct to deeper work, such as a warm-up (or a cool-down, strictly speaking) before meditation [Hack #60]. It's also useful as a means of gaining clarity and rationality when you're stressed or panicked, a sort of *emergency first aid* for clear thinking.

In Action

When you learn the Fourfold Breath, wear loose clothes or no clothes and make sure that you can breathe freely. Sit comfortably; you might want to lie down, but not if you're prone to falling asleep easily.

First, learn *belly breathing*:

1. Empty your lungs fully, until you can't empty them anymore.
2. Inhale slowly and deeply with the lower part of your lungs; it will feel as though you are breathing with your belly, from approximately the area of your navel. Only your belly should rise, not your chest.
3. Repeat. Belly breathing might take you some time to learn. Don't hurry it.

After you're comfortable with belly breathing, learn *chest breathing*: breathing with the upper part of your lungs only. Since this is the way most people in Western culture breathe anyway, learning chest breathing should not be difficult to do. When chest breathing, only your chest should rise, not your belly.

Next, learn to combine belly and chest breathing for a *full breath*:

1. When you breathe now, fill your entire lungs, first by filling the lower part of your lungs with belly breathing, and then by filling the upper part of your lungs with chest breathing.
2. When you exhale, empty your lungs fully.

Finally, learn to breathe rhythmically, in a fixed, repetitive pattern. This is the Fourfold Breath proper. The pattern I learned, and which I have found most effective, is 4-2-4-2, as follows:

1. Inhale with a full breath to the count of four.
2. Hold the full breath in your lungs to the count of two.
3. Completely empty your lungs to the count of four.
4. Hold your lungs empty to the count of two.
5. Repeat.

Depending on your body's rhythms, you might find other equivalent breath counts to be more effective, such as 6-3-6-3 or 8-4-8-4. Notice that these patterns are some variant of 2x-x-2x-x. Some teachers recommend entirely different rhythms, however, such as 4-4-4-4, or even a *threefold breath* such as 4-2-2 (inhale four, hold two, exhale two, repeat).[1] Feel free to experiment.

You might be able to find the right rhythm for breathing by synchronizing your breath with your heartbeat. If you do so, you'll know it; you'll become even calmer very quickly, and might sink into deep relaxation.

Everyone's body and mind are different, so experiment to find the best rhythm for you. Focus your attention on your breath and your counting, making sure that all your breaths are full and the count is steady. The Fourfold Breath is not a race; focusing on doing it slowly and steadily is part of what will help you relax.

In Real Life

If you learn to perform the Fourfold Breath invisibly, you will be able to hack your breathing inconspicuously whenever you feel angry, threatened, or anxious. For example, I have used it at the dentist's office, and I have used it to stay calm and friendly during job interviews.

If you train yourself to use the Fourfold Breath in stressful situations, you might find that it becomes a reflex. It was amusing to me that Frank Herbert described something similar in his science fiction novel *Dune*. Near the beginning of the novel, two characters are left in the desert to die. Both of them have undergone the mystical Bene Gesserit training, however, and they find that a calming breath pattern kicks in, allowing them to think clearly about their desperate situation. In this case, the fictional Bene Gesserit meditation exercises were definitely founded on fact.

Apart from inculcating a useful reflex that probably reduces the ravages of stress on your body, this hack might not produce lasting change by itself. The Fourfold Breath can be used to support deeper, more enduring techniques such as meditation [Hack #60] and REBT [Hack #57], but in itself it is mainly useful as first aid—although it is amazingly powerful first aid.

Just as you can quell a potentially fatal allergic reaction with a timely injection of adrenaline, so you can stem an unpleasant or even dangerous panic reaction with the Fourfold Breath. However, adrenaline won't cure your allergies (though immunotherapy might), and you shouldn't expect this exercise to cure you forever of any panic attacks you are experiencing.

On the other hand, I don't want to undervalue the Fourfold Breath either. If you do suffer from panic attacks, insomnia, or other stress-related ailments, just knowing you have a hack that can help you control them may lead you to feel peaceful, and thereby reduce your problem's frequency and severity a great deal. That in itself might be a life-changing experience.

End Notes

1. Henningsson, Ceci. 1994. "Pranayama in Three Easy Steps." *http://yogaclass.com/pranayam.html*. (A more detailed treatment of this pranayama technique, and one that is well worth reading.)

HACK #60 Meditate

Learn the basics of insight meditation, a much subtler and surer (but slower) technique for gaining clarity than the Fourfold Breath.

There are many kinds of meditation. The kind described in this hack is called *vipassana*, or *insight meditation*. Vipassana is the primary meditation technique of Theravada Buddhism, which you might think of as "orthodox" Buddhism; it is the sect that has remained closest to the teachings of Siddhartha Gautama from 2,500 years ago.

These brief instructions on how to meditate are intended to be useful for insight meditation only; if you wish to try another kind, you will have to consult another source.

The benefits of insight meditation include the following:

- Clearing your mind of distractions
- Experiencing better concentration and awareness
- Developing insight into how your mind works
- Hearing the quieter voices in your mind beneath the constant chatter
- Reducing the ravages of the fight-or-flight reaction on your body
- Gaining rest and respite, a "cool heart"

The ultimate goal of insight meditation is *nirvana* or *nibbana*. Most people suppose that this term means something like *heaven* or *eternal bliss*; actually, its literal meaning is something like *cooling off*, and it refers to eliminating or minimizing the suffering you feel.[1]

Naturally, if you're boiling over with hatred and rage, you can't think clearly. Other strong emotional states can distract you as well, as can constant internal monologue. Meditation aims to reduce how often you find yourself in those states.

In Action

Before you begin, don't worry if you can't meditate "perfectly," especially if you have never meditated before, or haven't meditated for a long time. Meditation requires discipline, but being too hard on yourself is contrary to its spirit. It's better to get back into practice by meditating "badly" or only for a few minutes at a time than to give up completely.

You can use meditation like a first-aid kit; indeed, it's often said that when you are too angry, frightened, or depressed to meditate, that's when you need it most. At those times, meditation can be useful to help you regain the equilibrium to make better decisions. However, you will obtain deeper, longer-lasting results if you use meditation like an exercise program rather than like first aid. If you do, you will find that the calm and clarity you obtain from meditation will become less a quick fix and more a part of your life and personality.

Here are some instructions for a simple meditation session:

1. Establish a comfortable upright posture, either in a chair, sitting cross-legged on a cushion, or in any other way that won't encourage your falling asleep.

2. You can prepare with a preliminary meditation on compassion[2] or use the Fourfold Breath [Hack #59] to still and regulate your breathing somewhat.

3. Focus on your breath. "Watch" it; feel it enter your nostrils and fill your lungs. Don't force yourself to breathe in a particular pattern; let your body lead the way as you might let a horse find the way home. Eventually, your breathing will become slow and calm.

4. Observe your thoughts as they drift by. Don't try to "catch" them or "hold" them; watch them as though you are watching clouds drift by in the sky. After a thought drifts past, return your attention to your breath

5. Continue meditating for at least 10 minutes. Using an alarm that makes a sound helps here; it's easier to relax into a meditative state if you're not tempted to look at your watch. You may wish to bow or make some other gesture to mark mentally the end of your meditation session.

Continue the practice periodically for a few days. If you decide to take it further, try to push yourself to meditate for 20 minutes, and then 30 minutes twice a day. Meditating at roughly the same time every day can help you establish the habit.

After meditation, you might find you have a calmer, clearer mind. If you don't, don't worry; you might have to meditate a while before it clicks. Experienced meditators often find that their meditation practice over many years is a series of such clicks, as their meditation deepens.

How It Works

The theory of insight meditation is that by watching your thoughts as calmly and neutrally as you can while not getting involved in them, you short-circuit the dreary, circular process of observing something, thinking about it, reacting emotionally to your thoughts, observing your emotional reaction, thinking about it, reacting emotionally again, and so on. If you meditate, eventually this cycle, which is like ripples in a pond, will die down naturally and your mind will become still and clear.

The purpose of focusing on your breath is to provide a neutral place for your attention when thoughts are not appearing in your mind. In later stages of meditation practice, you might use another focus, such as your walking feet.

How meditation works from a scientific standpoint is only beginning to be understood. It is known that the physiological effects typically include muscle relaxation, a slowing of heart rate, a lowering of blood pressure, slowing of breath, reduction of oxygen consumption, and an increase in alpha rhythms in the brain.[3,4] However, most practitioners of meditation will tell you there's more to it than a handful of physiological effects, no matter how beneficial. It's something you have to experience from the inside to understand.

In Real Life

The book *Happy to Burn*,[5] which I have found to be the most helpful introductory guide to insight meditation, outlines six phases in its meditation program. If you start with the first step and move slowly and methodically forward, you can expect to make significant progress:

1. *Basic relaxation*
   Relax and focus on your body's sensations.

2. *Meditation*
   Establish a comfortable upright posture. Focus on your breath as the *main object*. Observe your thoughts as they drift by. (Short instructions for this phase appear in the "In Action" section of this hack.)

3. *Concentration*
   Follow your breath. Silently say "in" and "out" on your in and out breaths. Silently note "sitting, sitting" as you observe yourself sitting.

4. *Mind noting*
   Note thoughts and sensations as they drift through your mind with the words "thinking, feeling, hearing, smelling, tasting, seeing." (Emotions should be noted as "feeling.")

5. Bringing meditation into everyday life
> Learn walking meditation, in which the main object is now not your breath, but your feet. Develop awareness by mind noting in daily life.

6. Developing the Observer
> The Observer is a detached, calm part of you. You can "court" it by noting your mind noting itself.

You can combine insight meditation with Rational Emotive Behavior Therapy [Hack #57] to further short-circuit your reactive cycles. If your meditation makes you aware of any particularly disturbing thoughts or feelings that don't go away after meditation, you can dispute them with rational self-analysis.

The relief that meditation can provide from the daily round can be exquisite, and the respite from the fight-or-flight reaction is probably good for your body, too. It's very much like a taking a small vacation every day, and then again, it's like coming home.

End Notes

1. Buddhadasa Bhikku. "Nibbana for Everyone." *http://www.suanmokkh.org/archive/nibbevry.htm.*

2. Universal loving-kindness meditation; *http://www.saigon.com/~anson/ebud/mfneng/mind9.htm.*

3. Murphy, Michael, and Steven Donovan. 2004. *The Physical and Psychological Effects of Meditation. http://www.noetic.org/research/medbiblio/index.htm.*

4. Benson, Herbert. 1975. *The Relaxation Response.* Morrow.

5. Wells, Roger. 1997. *Happy to Burn: Meditation to Energize Your Spirit.* Lothian Books. Available gratis at the author's web site: *http://www.users.tpg.com.au/sankhara/.*

See Also

- Another good book on insight meditation is *Mindfulness in Plain English* by Bhante Henepola Gunaratana (Wisdom Publications). An older edition of the book is available online: *http://www.saigon.com/~anson/ebud/mfneng/minddist.htm.*

Hypnotize Yourself

HACK #61

Self-hypnosis is a powerful self-motivation hack for short-term goals. Contrary to popular opinion, it is not at all spooky or mysterious, and can be very effective.

Hypnosis is an altered state of consciousness that is available to most human beings without the use of drugs. Definitions of hypnosis vary, but most of them emphasize the following characteristics of a hypnotic trance:

- Relaxation
- Concentration
- Suggestibility

This hack will help you motivate yourself, boost your confidence, and achieve a short-term goal. With regular use, self-hypnosis may also allow you to achieve longer-term goals and change other things about your thinking patterns as well. It is quite safe for most people, and since you are hypnotizing yourself in a quiet, isolated environment, you need not fear that anyone will make you think you're a chicken.

> Research varies as to whether a person can actually be made to do something under hypnosis against her will, although most professionals believe this is unlikely. Nevertheless, if you are undergoing psychological counseling or you are concerned about your mental state, you should consult your mental health care professional to determine whether this procedure is advisable for you.

In Action

Self-hypnosis is a reasonably simple process, but you should take the time to plan. The following process is a basic outline for a self-hypnosis session.[1] Read over all the steps before you start, assemble whatever you need to be comfortable, and give it a little thought. Your effort will be rewarded.

Set your goals. You'll need to think carefully about what messages you wish to send yourself while under hypnosis. Hadley and Staudacher[2] give some basic ideas about how to formulate effective hypnotic suggestions:

They should be worded simply and repeated several times.
    It's best to focus on one change at a time and state it simply in a way that will be easy to repeat and remember.

They should be believable, obtainable, and desirable.
Use self-hypnosis to work on plausible goals, things that are possible for you to do and that you can imagine doing, even if you need a little help. You should break larger goals into achievable steps so that you can work with them one at a time. Also, it might go without saying, but giving yourself goals that you really want will help you to achieve them.

They should be stated positively and for a specific time.
There are two sides to every coin, so choose to reinforce the positive side of what you want to achieve rather than the negative. In other words, choose to say, "I will be relaxed" rather than "I won't be anxious." Further, give yourself specific time parameters when applicable, such as "I will work out from 6:30 to 7:30 tonight."

They should include cue words or a key phrase that will trigger your desired reaction.
For example, give yourself the cue word "relax" if you're using hypnosis to help yourself be calm during a job interview, or the phrase "go for it" if you want to make yourself enthusiastic about a workout.

They should provide detailed images of the desired outcome.
If you're using hypnosis to calm yourself for a job interview, be ready to tell yourself what that's like: "I am relaxed and confident...I am answering questions articulately...the interviewer is impressed with my poise..." and so on.

Thinking about these pointers, formulate the suggestion you will give yourself. For example, if your nervousness that you can pass a test is interfering with your ability to study, you can create a suggestion such as, "I know that I can and will pass this test."

Note that you are telling yourself three things in this short suggestion: you *can* pass the test (you have the ability), you *will* pass the test (you have the determination), and you *know* both of these things (you have the confidence). If you are a programmer, you can think of a suggestion as an elegant code hack.

See how much information you can pack into a single short suggestion. Giving your suggestion a rhythm or cadence ("I *know* that I *can* and I *will* pass this *test*") can help the suggestion to stick. Add your image about your desired outcome, and you're ready to begin.

Lie down. The room should be dim or darkened. You should wear comfortable, loose-fitting clothes or be naked.

Induce self-hypnosis. Tell yourself, "As I count backward from 100 to 0, I will descend deeper and deeper into hypnosis. The deeper I go, the more suggestible my unconscious mind will be. When I reach 0, my unconscious mind will be completely open to the suggestion I will give it. After I give myself the suggestion, I will awake, completely refreshed and ready."

Count backward from 100 to 0. You might imagine yourself descending a staircase into the ground, with a landing every 10 steps to keep yourself from mentally leaping too far ahead. When you reach 0, you can imagine yourself emerging into a large, open cavern. There's no reason it needs to be a cavern, however; you might imagine yourself emerging into a sunny wonderland or onto a cliff overlooking the sea—or anywhere else that you feel comfortable, safe, and powerful.

Test your depth of hypnosis. Tell yourself, "I am now deep in hypnosis. My arm is heavier than lead. I will try to lift my arm, but I will not be able to do so because it is too heavy." If you are sufficiently deep in trance, you will find that you cannot in fact lift your arm.

Don't worry about whether you're just "pretending"; if you can't lift your arm, it will be an unmistakably peculiar sensation! If you find you can lift your arm, that's all right, too. You're new at this; give yourself another chance by starting again to induce hypnosis.

Give yourself the suggestion. You might want to repeat the suggestion (for example, "I know that I can and I will pass this test") a predetermined number of times, such as 20, or you might want to repeat it until you sense that it has *clicked* and further repetition will not strengthen it.

Allow yourself to emerge from the trance. Tell yourself, "I have done what I set out to do in this session. I am confident that my suggestion has taken. At the count of three, my eyes will spring open, and I will awaken from the trance, completely refreshed and ready to implement my suggestion. One...two...three!" You should now awake.

 You might find it useful to record your self-hypnosis script on an audiocassette or MP3 player and to follow the recording instead of trying to keep the instructions straight in your head as you enter the trance.

How It Works

Everyone has had the experience of seeing a great movie and really sinking into it. When you come out of the theater, you might feel disoriented; it's hard to believe that you weren't in the movie, and you're emotionally drained because you've had the emotional involvement and reactions as if you were there. Basically, self-hypnosis allows you to run a movie in your mind. With practice, you can project the movie as well as get involved with it, so you can decide what movie you want to watch and what emotions and beliefs you want to elicit from yourself.

Self-hypnosis works, as the great movie does, because it allows us to see situations as almost real, with the safety of knowing that it isn't really real. We are changed by going through the emotional reactions and experiences as if they were real. Self-hypnosis also gives us an opportunity for goal rehearsal, which allows us to imagine what having a goal would be like. By doing this, we shape our minds so that we know what it will be like to achieve the goal, which helps us know how to get there from where we are and puts us into the mindset to look for opportunities and expect success.[3]

What can you do with self-hypnosis besides become more confident? Many claims have been made for hypnosis, although some of them are controversial. In clinical testing, self-hypnosis has been shown effective in reducing pain, even in patients that had been very resistant to other pain-reduction methods[4] and patients with chronic pain.[5] In another study, subjects increased their physical exercise performance by using hypnosis.[6] Other claims include breaking habits such as smoking, overcoming phobias, recalling forgotten information, directly controlling physiological processes such as bleeding, and eliminating infections such as the virus that causes warts. Readers of *Mind Performance Hacks* might also find it interesting that claims have been made that hypnosis can increase reading speed and the ability to do mental math.

Skeptics debate many of these claims, including the one that hypnosis has been shown reliable at recovering lost or repressed memories. In fact, hypnosis can be used to create *false memories*, and memories "recovered" under hypnosis are therefore not admissible as evidence in many courts of law. *The Skeptic's Dictionary* claims (emphasis ours):

> When one strips away [the] dramatic dressings [of hypnosis], what is left is something quite ordinary, *even if extraordinarily useful*: a self-induced, "psyched-up" state of suggestibility.[7]

This hack was written with the understanding that self-hypnosis can, at the very least, enhance your ability to perform anything you could normally do in a relaxed, focused, psyched-up, open-minded mental state, and that it's quicker than an all-day motivational seminar, as well as a lot cheaper.

In Real Life

A self-hypnosis session such as the one described in the previous section should be enough to get you through a hard weekend study session or something of similar intensity and duration. In general, self-hypnosis works best for short-term goals.

However, you can use it for long-term goals as well, if you keep refreshing your suggestion. In this way, it is like meditation [Hack #60], which is also important to do on a regular basis if you want anything more than short-term results. Marty has used self-hypnosis techniques (as well as the Four-fold Breath [Hack #59]) for years to tackle chronic anxiety attacks. She uses a couple of different hypnosis techniques; some help avoid attacks, and others help her stop an attack in progress. She's had great success with this, both in the short-term goal of stopping and escaping from anxiety attacks and in the longer-term goal of avoiding them.

Even a short-term effect can be useful if it gets you over a hurdle, and short-term effects can build into long-term effects. For example, passing that test might give you such confidence in the future that you won't *need* to renew the suggestion. Marty has used her anxiety hypnosis so effectively in the moment that it has dramatically reduced her likelihood of suffering attacks over time.

End Notes

1. Copelan, Rachel. 1984. *How to Hypnotize Yourself & Others*. Bell Publishing Company.

2. Hadley, J., and C. Staudacher. 1985. *Hypnosis for Change*. New Harbinger Publications.

3. Soskis, D. 1986. *Teaching Self-Hypnosis: An Introductory Guide for Clinicians*. W. W. Norton & Company.

4. Gutfeld, G., and L. Rao. 1992. "Use of Hypnosis with Patients Suffering from Chronic Headaches, Seriously Resistant to Other Treatment," as reported in *Prevention*, 44: 24–25.

5. Barabasz, A.J., and M. Barabasz. 1989. "Effects of Restricted Environmental Stimulation: Enhancement of Hypnotizability for Experimental

and Chronic Pain Control." *International Journal of Clinical and Experimental Hypnosis*, 37: 217–231.

6. Jackson, J.A., G.C. Gass, and E.M. Camp. 1979. "The Relationship Between PostHypnotic Suggestion and Endurance in Physically Trained Subjects." *International Journal of Clinical and Experimental Hypnosis*, 27: 278–293.

7. *The Skeptic's Dictionary*. "Hypnosis." *http://skepdic.com/hypnosis.html*.

—*Ron Hale-Evans and Marty Hale-Evans*

HACK #62 Talk to Yourself

Language isn't just for talking to other people; it may play a vital role in helping your brain combine information from different modules.

Language might be an astoundingly efficient way of getting information into your head from the outside, but that's not its only job. It also helps you think. Far from being a sign of madness, talking to yourself is something at the essence of being human.

Instead of dwelling on the evolution of language and its role in rewiring the brain into its modern form[1], let's look at one way the brain can use language to do cognitive work. Specifically we're talking about the ability of language to combine information in ordered structures—in a word: syntax.

Peter Carruthers, at the University of Maryland,[2] has proposed that language syntax is used to combine, simultaneously, information from different cognitive modules. By "modules," he means specialized processes into which we have no insight,[3] such as color perception or instant numbers. You don't know *how* you know that something is red or that there are two coffee cups, you just *know*. Without language syntax, the claim is, we can't combine this information.

The theory seems pretty bold—or maybe even wrong—but we'll go through the evidence Carruthers uses and the details of exactly what he means, and you can make up your own mind. If he's right, the implications are profound, and it clarifies exactly how deeply language is entwined with thought. At the very least, we hope to convince you that *something* interesting is going on in these experiments.

In Action

The experiment described here was done in the lab of Elizabeth Spelke.[4] You can potentially do it in your own home, but be prepared to build some large props and to get dizzy.

Imagine a room like the one in Figure 7-1. The room is made up of four curtains, used to create four walls in a rectangle, defined by two types of information: geometric (two short walls and two long walls) and color information (one red wall).

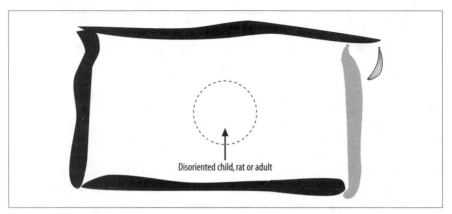

Figure 7-1. Setup for Spelke's experiments—a rectangular room with one colored wall

Now, think about the corners. If you are using only geometric information, pairs of corners are identical. There are two corners with a short wall on the left and a long wall on the right, and two corners the other way around. If you are using only color information, there are also two pairs of identical corners: corners next to a red wall and corners *not* next to a red wall.

Using just one kind of information, geometry or color, lets you identify corners with only 50% accuracy. But using both kinds of information in combination lets you identify any of the four corners with 100% accuracy, because although both kinds of information are ambiguous, they are not ambiguous in the same way.

So, here's a test to see if people can use both kinds of information in combination.[5] Show a person something he'd like, such as some food, and let him see you hide it behind the curtains in one corner of the room. Now disorient him by spinning him around, and ask him to find the food. If he can combine the geometric and the color information, he'll have no problem finding the food—he'll be able to tell unambiguously in which corner it was hidden. If he doesn't combine information across modules, he will get it right 50% of the time, and he will get it wrong 50% of the time on his first guess and need a second guess to find the food.

Where does language come into it? Well, language seems to define the kinds of subjects who can do this task at better than 50% accuracy. Rats can't do it. Children who don't have language yet can't do it. Postlinguistic children and adults can do it.

Convinced? Here's the rub: if you tie up an adult's language ability, her performance drops to close to 50%. This is what Linda Hermer-Vazquez, Elizabeth Spelke, and Alla Katsnelson did.[6] They got subjects to do the experiment, but all the time they were doing it, they were asked to repeat the text of newspaper articles that were played to them over loudspeakers. This "verbal shadowing task" completely engaged their language ability, removing their inner monologue.

The same subjects could orient themselves and find the correct corner fine when they weren't doing the task. They could do it when they were doing an equivalently difficult task that didn't tie up their language ability (copying a sequence of rhythms by clapping). But they couldn't do it with their language resources engaged in something else. There's something special about language that is essential for reorienting yourself using both kinds of information available in the room.

How It Works

Peter Carruthers thinks that you get this effect because language is essential for conjoining information from different modules. Specifically he thinks that it is needed at the interface between beliefs, desires, and planning. Combining across modalities is possible without language for simple actions, but there's something about planning, and that includes reorientation, which requires language.

This would explain why people sometimes begin to talk to themselves—to instruct themselves out loud—during especially difficult tasks. Children use self-instruction as a normal part of their development to help them carry out things they find difficult.[7] Telling them to keep quiet is unfair and probably makes it harder for them to finish what they are doing.

If Carruthers is right, it means two things. First, if you are asking people to engage in goal-oriented reasoning, particularly if it uses information of different sorts, you shouldn't ask them to do something else that is verbal, either listening or speaking.

 I've just realized that this could be another part of the reason people can drive with the radio on but need to turn it off as soon as they don't know where they are going and need to think about which direction to take. It also explains why you should keep quiet when the driver is trying to figure out where to go next.

Second, if you do want to get people to do complex multisequence tasks, they might find it easier if they can perform the tasks using only one kind of information so that language isn't required to combine across modules.

End Notes

1. Although if you do want to dwell on the role of language in brain evolution (and vice versa), you should start by reading Terrence Deacon's fantastic *The Symbolic Species: The co-evolution of language and the brain* (W. W. Norton & Company).

2. The article containing this theory was published by Peter Carruthers in *Behavioural and Brain Sciences*. It and the response to comments on it can be found at *http://www.philosophy.umd.edu/people/faculty/pcarruthers/ Cognitive-language.htm* and *http://www.philosophy.umd.edu/people/faculty/ pcarruthers/BBS-reply.htm*.

3. OK, by "modules," he means a lot more than that, but that's the basic idea. Read Jerry Fodor's *The Modularity of Mind* (MIT Press) for the original articulation of this concept. The importance of modularity is also emphasized by evolutionary psychologists, such as Steven Pinker.

4. Much of the work Peter Carruthers bases his theory on was done at the lab of Elizabeth Spelke (*http://www.wjh.harvard.edu/~lds*).

5. Strictly, you don't have to use both kinds of information in combination at the same time to pass this test; you could use the geometric information and then use the color information, but there is other good evidence that the subjects of the experiments described here—rats, children, and adults—don't do this.

6. Hermer-Vazquez, L., E. S. Spelke, and A. S. Katsnelson. 1999. "Sources of flexibility in human cognition: Dual-task studies of space and language." *Cognitive Psychology*, 39(1): 3–36.

7. Berk, L. E. November 1994. "Why children talk to themselves." *Scientific American*, 78–83. *http://www.abacon.com/berk/ica/research.html*.

—Matt Webb and Tom Stafford

Interview Yourself

Use interviewing techniques as part of your search strategy for solutions to problems. In particular, understand how interviewers use open and closed questions, and use this knowledge to keep your own search on track.

Interviewers have a range of well-honed techniques that they use to gather information. You can apply the psychology of interviewing to yourself to gain perspective on your own thinking, particularly where you need to make a decision.

In Action

Start with an objective, a choice or decision that you need to make, in which the right decision is far from clear. A decision that has many competing factors to consider is ideal. This decision is to be the focus of the exercise. You'll be gathering information from yourself that will help you make that decision. It's a bit tricky, because you'll be playing the role of both interviewer and interviewee.

Find somewhere comfortable enough to sit, but not the most comfortable chair in the house, and have a pad of paper to make notes on. If you've already collected some written information, quickly scan through it to remind yourself of what you already have.

Start out with easy questions to which you know the answers and, ideally, feel good about. The questions need to be related to the overall objective, but the purpose of asking them is to *warm up* and get answers coming easily.

As you get into this, gradually switch to deeper questions. The questions should be prompting you to think about different aspects of the decision. Remember, you're gathering information about the decision, not actually making the decision at this point. Resist the temptation to *approve* or *disapprove* of an answer. That can get in the way of gathering the information, whereas maintaining your *journalistic objectivity* can help.

As an interviewer, the question in your mind as you listen to each answer should be "Do I need to know more in that direction?" As you're doing this, notice whether you're making useful progress. Since you're both interviewer and interviewee, the process should be more of a partnership than an adversarial game—in other words, more like a typical newspaper interview than a typical job selection interview. Nevertheless, as interviewer, keep your wits about you and make sure your interviewee isn't dodging any of the questions.

The postponement of evaluation used in this hack is related to *brainstorming*, a technique used in groups to promote creativity. In brainstorming, many ideas are encouraged and none is criticized. The ideas are then evaluated as a separate phase at the end of the process.

Here is the crucial part. If you find you're moving away from the key objective, use more focused questions, known as *closed questions*, for which the answers are a simple "yes," "no," "I don't know," number, date, or name. When you ask a closed question, you need to be ready to follow up with more questions, since you're looking for a short answer.

Conversely, if you feel something is being left unsaid, or some aspect of the decision is not being considered, go for more *open questions*, questions that give more opportunity for more expansive—and more surprising—answers. It's perfectly possible to surprise yourself with the answers you might find. The key is a balance between the open and closed questions. In your role as interviewer, you must actively adjust that balance. You want to keep getting information, but ensure that it stays sufficiently on-topic and relevant to your overall task.

Give yourself a time limit—say, 40 minutes. At the end of that time, bring the interview process to a clean conclusion, take a short break, and then review what you've learned from the process.

The ritual described is a bit artificial, and you can gain much of its benefit more simply after some experience by recognizing in your own thinking what subgoals you are setting for yourself or what questions you are asking yourself. More creativity will come with the open questions, more focus with the closed ones. The formal interview process, however, will help you develop and strengthen these skills.

> You can increase the possibility of taking yourself by surprise by visualizing yourself confronted by an interviewer or interviewee. This is less awkward than doing the same thing physically, by scampering back and forth between two chairs. You can also merge this hack with "Adopt a Hero" [Hack #31]: imagine yourself being interviewed by an appropriate person, or imagine interviewing them.

In Real Life

Here is an approximation to an interview with myself. I wrote it from the interviewer's perspective, as if it were between two people and spoken out loud. In actuality, it was unspoken and I was both the interviewer and interviewee. The problem I chose to explore in the interview was a challenging one—coming up with a birthday present for my sister.

> Tell me some things about your sister. What's she like?

An open question, and not too difficult. A good place to start. However, there was a problem straight away. The interviewee quickly shifted to talking about things his sister isn't interested in, implying that this makes finding a present more difficult. To get him back on track:

> And what interests her? What do you talk about with her?

Prompted to mention specific interests, the interviewee fairly rapidly came up with five general areas.

> Do any of these provide good opportunities for presents?

More problems. The interviewee's response to the question was noncommittal. He seemed reluctant to engage with the question. Perhaps the question was too open and needed a tighter focus. One option was to now home in on just one of the interests and see where that led. Instead of narrowing the search so much so soon, I instead tried:

> How about something that combines two of the five interests that you've identified?

With some more encouragement, this led to a mix of "off the wall" ideas, and some that might be promising. Instead of evaluating here and now whether the ideas were good or bad, the follow-up questions I asked were to gather more information about these ideas. These were mostly closed questions, questions about where to buy, type, color, approximate cost—questions about practicalities. However, these didn't turn out quite as closed as I intended. In particular, the "where to buy" question led the interviewee to a similar, and in his opinion slightly better, idea for one of the possible gifts. Now, noticing that time was nearly up, I asked the interviewee whether he wanted to add anything to what he'd said already. He didn't, so I brought the interview to a close.

At this point, I took a brief break and then I came back and went through the ideas, junking the nonstarters. That left just four. Of these, two were on the expensive side, too risky to get for her without asking her first whether she'd like them, but they were definite contenders. One of the ideas I hadn't a clue where to get: "Try eBay, perhaps?" The fourth was a bit on the mingy side. As I'd have been pleased to get one decent idea, I thought that was a good and useful result.

How It Works

The hack works because interviewers have refined the art of interviewing through a great deal of experience. Putting someone at ease at the start of an interview is essential to getting good information from him. Interviewers also know from experience that they stop getting accurate information from a candidate when they show strong approval or disapproval of the candidate's responses. The most verbal parts of thinking appear to be an abbreviated form of conversation, so similar principles apply.

The key part of the hack is having a search strategy and using it consciously. Open questions widen the search area, and closed questions home in on details to keep the search on track.

See Also

- Malcolm Peel's book *Readymade Interview Questions* (Kogan Page) provides a well-structured introduction to the interviewing process.
- *Interviewing* by Glynis M. Breakwell (Routledge) covers media and research interviews as well as selection interviews, but is much less focused on the actual questions to ask than Peel's book.

—James Crook

H A C K   Cultivate the Naive Mind
#64

It has been said that the human computers in Frank Herbert's Dune novels cultivated "the naive mind" to process information without bias. The guidelines of the Wikipedia neutral point of view (NPOV) are a tested way of cultivating the naive mind in your thought and communication.

The completely objective person is a common trope in science fiction, from the *mentats* (so-called "human computers") in *Dune*, to the legal profession of Fair Witness in Robert Heinlein's novel *Stranger in a Strange Land*, to the character of Mr. Spock on *Star Trek*. Being dispassionate is also a fairly common geek ideal, and occurs in both Western and Eastern philosophies (the Stoic Sage and the Buddha). However, some postmodern philosophers such as Paul Feyerabend and Richard Rorty have questioned the concept of objectivity, saying that every human being is biased in some way. "You can't be objective" has become a commonplace in certain segments of our culture, such as academia.

The Wikipedia Project, a collaborative web-based encyclopedia that has surpassed the *Encyclopaedia Britannica* both in the number of articles it contains and in the amount of Internet traffic it receives, recognizes the ubiquity of human bias and the hard problem of objectivity, but has attempted to forge a practical official policy anyway. The project has developed a set of guidelines it calls the *neutral point of view* (NPOV).

Despite the fact that any casual reader can click the "edit this page" button on a Wikipedia article to fill it with bigotry and nonsense (for any values of "bigotry" and "nonsense" you care to define), the NPOV enables Wikipedians to reach a stable consensus on articles ranging from capitalism to abortion, to the death penalty. These guidelines can be a useful mental-clarity hack because they can enable you to settle arguments, see other points of view, reveal your *own* invisible biases, and, most importantly, transmute the dung of biased human opinion into the gold of uncontested fact.

In Action

The neutral point of view, as Wikipedians define it, has a very distinct meaning. According to this standard, *neutral* and *unbiased* do not mean the same thing as *objective*. Objectivity implies a kind of God's-eye view, which is impossible for finite intelligences, whereas the neutral point of view simply means not taking sides (however hard that might be). Here's a brief summary of the official Wikipedia policy:

> Unbiased writing does not present only the most popular view; it does not *assert* the most popular view is correct after presenting all views; it does not assert that some sort of intermediate view among the different views is the correct one. Presenting all points of view says, more or less, that *p*-ists believe that *p*, and *q*-ists believe that *q*, and that's where the debate stands at present. Ideally, presenting all points of view also gives a great deal of background on who believes that *p* and *q* and why, and which view is more popular (being careful not to associate *popularity* with *correctness*). Detailed articles might also contain the mutual evaluations of the *p*-ists and the *q*-ists, allowing each side to give its "best shot" at the other, but studiously refraining from saying who won the exchange.[1]

Let's try to clarify the basis for the neutral point of view.[2] The NPOV attempts to state facts and only facts. A *fact* is a statement about which no serious dispute exists, such as "Saturn is a planet" or "Submarines can dive underwater." On the other hand, an *opinion* or *value* is a statement about which some dispute exists, such as "Writing graffiti on walls is wrong," or "*The Lord of the Rings* is the best book of the 20th century."

If you wish to communicate from the neutral point of view, *describe* disputes, don't *engage* in them, even if you agree with one side. (This might be thought of as the NPOV golden rule.) For example, instead of saying, "*The Lord of the Rings* is the best book of the 20th century," which is a value or opinion, say, "In a 1997 survey by the British bookseller Waterstone's, *The Lord of the Rings* was voted the best novel in English of the 20th century, capturing about 5,000 votes from the 25,000 people surveyed." This is a fact; you have converted an opinion into a fact.

In a similar way, you can convert other values and opinions to facts by attributing them to identifiable spokespeople or populations. When you do, be sure to explain the reasons why the people who hold those opinions do so, make clear which is the majority view and which are the minority views, and don't imply that one side or the other is correct. Remember, "Writing graffiti on walls is wrong" is an opinion, and "78% of Americans agree with the U.S. Anti-Graffiti League that writing on walls that don't belong to you is wrong, 12% believe it to be 'a righteous act of civil disobedience and self-expression,' and 10% state they 'couldn't care less'" is a fact.

At least, it's *factlike*. I invented the organization and the statistics for the sake of an illustration that I thought most people could read calmly.

Finally, describe opinions in a positive, sympathetic way, especially if you disagree with them. Who knows? You might be wrong, and you'll be glad one day when someone who disagrees with *your* minority opinion treats it respectfully.

Leaving aside the Wikipedia, if you apply these principles to the thought and communication in your own life, you may find yourself becoming a calmer, less contentious, and more understanding person. Is learning the NPOV guidelines worth the effort? I claim it is, and that's a fact.

In Real Life

Here is an example of a biased text made neutral by application of the NPOV. The following text is excerpted from the home page of AGHOST (the Amateur Ghost Hunters Of Seattle/Tacoma), which you can find at *http://www.aghost.us*:

> A.G.H.O.S.T. is the most advanced technical paranormal research group in the Pacific Northwest. During paranormal investigations, our team combines hi-tech equipment with qualified psychics, research and training for successful results. We work with the basic investigative tools such as cameras (35 mm and digital), audio and video recording devices, and EMF detectors, as well as the most advanced computerized surveillance equipment including infrared motion sensor cameras. Each member of our investigation team has been thoroughly trained in their areas of responsibility. AGHOST has earned the respect of some of the most important names in the paranormal field.

> We, the ghost hunters, have made it our quest to provide substantial evidence that we are not alone in the dark. What is it that goes bump in the night and gives you the feeling of being watched when you're all alone? With the work of ghost hunters today...the TRUTH will be found.

Here's the text after applying the NPOV guidelines to it:

> A.G.H.O.S.T. is a research group in the Pacific Northwest that investigates phenomena we suspect have a paranormal origin. Our investigatory team uses equipment such as 35mm and digital cameras, audio and video recorders, electromagnetic field detectors, and Geiger counters, as well as surveillance equipment including laptops with infrared motion sensor cameras and other sensors.

> Our team also includes members who claim to have psychic abilities, and we listen seriously to what they tell us. We train all members of our investigation team in their special areas of responsibility, and the results they obtain

sometimes surprise the journalists and skeptics who accompany us on expeditions. AGHOST has also earned the respect of such prestigious investigators of the paranormal as Pam Psychic, who said, "AGHOST has shown positive proof of more ghosts than any other amateur ghost-hunting group I know of," and Sam Skeptic, who said, "Although I don't believe in ghosts myself, I respect AGHOST. They don't jump to conclusions."

Although many of the ghost hunters of AGHOST believe in life after death, we recognize that not everyone does. However, even skeptics experience an eerie feeling from time to time when they hear a bump in the night or feel as though they're being watched when they're alone. As ghost hunters armed with twenty-first-century technology and a team we trust, we believe it is our duty to find the truth behind mysterious occurrences.

The altered version of the AGHOST promotional material relies more on facts than opinions. Should AGHOST ever adopt such an approach on its home page, in my estimation, the loss of "true believers" who would find the more neutral approach repugnant would be more than offset by the influx of skeptics and merely curious people who would find it refreshing.

 If you're attempting to persuade other people dishonestly, you might be able to develop a kind of pseudoneutrality that looks neutral on the surface but actually stacks the deck by citing facts and opinions that support what you want to prove and passing over evidence that supports another conclusion. However, if you set out to use this hack as a mental training technique and you don't make an honest effort at real neutrality, you're cheating only yourself.

End Notes

1. Wikipedia. 2005. "Wikipedia: Neutral point of view." *http://en.wikipedia. org/wiki/Npov.*

2. Wikipedia. 2005. "Wikipedia: NPOV tutorial." *http://en.wikipedia.org/ wiki/Wikipedia:NPOV_tutorial.*

See Also

- What goes wrong when NPOV guidelines are *not* followed: *http://en. wikipediaorg/wiki/Wikipedia:Lamest_edit_wars_ever.*
- The science-fictional concept of "the naive mind" might have originated in part with the Zen concept of "beginner's mind" (a mind without preconceptions), as described by Shunryu Suzuki in his book, *Zen Mind, Beginner's Mind* (Weatherhill).

HACK #65 Employ Mental Momentum

Recognize your mind's tendency to keep doing what it's doing, and you might learn to see the forest as well as the trees, for your own benefit.

Your mind is like your car: it has its own direction and momentum. The momentum is important: once your mind is moving in a certain direction (that is, focused on a certain subject), it will continue moving in that direction until something—either the driver (you) or a concrete embankment (life)—alters its course.

For our purposes, mental momentum comes in two varieties: *positive* and *negative*. Positive mental momentum happens when you are caught up in something you ought to be doing; negative mental momentum happens when you are caught up in something you ought not to be doing.

Productivity consultant Alan Lakein developed something he called the *Swiss cheese method*: breaking up a large project into many small tasks of five minutes or less[1] (this is analogous to the *next task* concept in the *Getting Things Done* system[2]). If you start these tasks with the intention of working on them for only five minutes, not only will you drill a lot of five-minute holes in the cheese of your project, but also—and this is important—you might find that you don't want to stop when the five minutes are up. This is positive mental momentum.

Unfortunately, mental momentum also comes in the negative variety. Unless you're aware of it—and your own predilection toward it—negative mental momentum can prevent intelligent focus. Negative mental momentum is the proverbial human tendency of not seeing the forest for the trees—or becoming entranced with the patterns in the bark of one particular tree of the forest, or the colors of the carapace of a beetle on the bark.... The tendency to engage in small and manageable but irrelevant activities with nice boundaries, and not get around to less well-defined but more important activities, is endless and comes in many forms:

- Getting caught up in a fun waste of time when you have important work to do

- Getting stuck on fixing one small bug in a program when many other bugs need to be fixed (some of which might fix the current bug as a side effect)

- In a board game such as chess or Go, defending one piece or one small area of the board fiercely while your opponent quietly sets up traps for other pieces or board areas

Chess Grandmaster Yasser Seirawan used to be known for tricks such as setting up a queen-side attack as a decoy and then slamming his opponent with a surprise attack on the king side.

- Strenuously defending a logical position even if it wasn't thought out well in the first place, or was said only in jest

Many of these kinds of mental and emotional momentum seem to be leftovers from our brutal evolutionary past; the more primitive mammalian and reptilian parts of our brain are wired to defend territory to the death, and some part of us thinks of a chess position, or a logical position, as territory.[3]

According to relativity theory, depending on your frame of reference, momentum is the same as inertia, so, reasoning analogically [Hack #25], if you're doing nothing special (mental inertia), you'll probably continue to do so. We can see that this is true in real life; people in this situation are said to be *drifting*.

Related to the concept of mental momentum is the concept of long-range hedonism put forth by Albert Ellis, developer of Rational Emotive Behavior Therapy [Hack #57]. It's normal for humans to desire pleasure and wish to avoid pain, but too many of us sacrifice long-term pleasures (such as obtaining an advanced degree and its perquisites) for short term pleasures (such as taking "study breaks" that turn into all-night parties). Although there's nothing wrong with short-term pleasures as such, it is often rational to prefer long-term pleasures to short-term pleasures and to tend to sacrifice the short term for the long term, instead of the other way around.[4]

It's not *always* rational! If you defer your pleasures long enough, you'll be dead.

In Action

Make a plan and stick to it. Periodically remind yourself somehow, such as with an exoself [Hack #17], of what you need to be doing. Planning from the top down (from the forest to the trees in it) and not from the bottom up (from the trees to the forest) is important. In terms of the *Getting Things Done* system, asking "What is the next task?" should flow from "What am I trying to accomplish?" instead of the other way around.

It is crucial to assess how much of a tendency you have toward both positive and negative momentum. If you have a strong tendency toward negative momentum, avoid doing anything that isn't part of your master plan and don't spend too much time on small tasks, planned or not. On the other hand, if you find you have a tendency toward positive momentum, make maximal use of the Swiss cheese method described earlier.

In Real Life

Here is a story about putting the brakes on some negative mental momentum. I recently bought a used laser printer that didn't have enough memory to print duplex and didn't come with PostScript installed. I ordered some more memory for the printer from eBay, but meanwhile, I had a 200-page PDF file that I wanted to print four-up and duplex. I woke up one day with the intention of writing hacks for this book, but decided I would "just" print the file duplex manually. I thought I knew the right PostScript utilities on my GNU/Linux box to create a four-up PostScript file that I could then print duplex by splitting it into front and back pages. Of course, my various PDF, PostScript, and GhostScript utilities didn't want to talk to one another, so I had to learn about some PostScript cleanup and compatibility utilities, and when those didn't work right....

Suddenly, it was 6 p.m. I looked up and said to myself, "This is ridiculous! You've done it again!" I had become entranced by the beetle on the bark. Fortunately, I was able to remind myself of my goals for the day and to shake off the evil enchantment under which I had allowed myself to fall. The situation even crystallized some thinking I had been doing about the concept of positive and negative mental momentum (leading to this hack), so not all was lost.

End Notes

1. Lakein, Alan. 1973. *How to Get Control of Your Time and Your Life*. Signet. This book was to the 1970s what *Getting Things Done* is to the 2000s.

2. Allen, David. 2003. *Getting Things Done: The Art of Stress-Free Productivity*. Penguin Books.

3. Wilson, Robert Anton. 1983. *Prometheus Rising*. Falcon Press. Flaky, but has some remarkably perspicacious things to say about the human "territorial" attachment to ideas.

4. Ellis, Albert, Ph.D., and William J. Knaus, Ed.D. 1977. *Overcoming Procrastination*. Penguin Books USA. The whole book is written in E-Prime [Hack #52]!

CHAPTER EIGHT

Mental Fitness
Hacks 66–75

Physical fitness is commonly defined as consisting of strength, flexibility, and endurance. You can also think of *mental fitness* as consisting of these characteristics. Mental strength is the ability to attack a problem, mental flexibility is the ability to stretch your mind to see all of the problem's aspects, and mental endurance is the ability to keep at a problem until you solve it.

This last chapter puts together everything you learned in the previous chapters. It contains hacks about caring for your brain with sleep, nutrition, and exercise [Hack #69], and mental warm-ups [Hack #66] and mind sports [Hack #67]. Finally, the last hack in the book enables you to integrate all the previous hacks and carry them around in a big, red, mental toolbox [Hack #75] that you can bring with you, even where there is no electricity or pen and paper.

HACK #66 Warm Up Your Brain
Get the blood flowing to your gray matter with a few mental push-ups.

Sometimes the mind simply doesn't want to think. The morning caffeine hasn't kicked in yet, or perhaps it is wearing off. Thought seems like an effort. In this situation, you might normally be inclined to wait until the sluggishness wears off, perhaps by taking a break with someone else's thoughts (browsing the Web, reading a book, or watching TV, for example).

If you would really rather get busy, you might be able to push past the threshold of resistance by doing a little mental warm-up. If thinking is normally a pleasurable activity for you, a small amount of it that requires only a small effort can remind you of this pleasure and motivate you to begin or resume your thinking in earnest.

A good mental warm-up activity is one that is small in scale and gives you at least a small sense of accomplishment. The smallness of scale is important so that the warm-up is easy to begin and easy to end. Ideally, the warm-up will require no preparation so that you can start on a whim.

You must also be able to stop on a whim. Although this might seem like a fast path to abandoning the warm-up in exchange for the web browser, it is important that you be able to stop once you feel mentally alert. This might be difficult if you get too much pleasure out of the exercise or if it involves a goal that you end up wanting to achieve. For this reason, small puzzles (for example, brainteasers) are good for mental warm-ups, but large puzzles are not. A large puzzle (for example, a crossword puzzle) might warm you up (a crossword puzzle is composed of many small puzzles, after all), but it might also suck you in and not release you until you've completed the whole thing.

In Action

Here's a mental warm-up exercise that requires no preparation:

1. Pick a small (one- or two-digit) number.
2. Double it in your head.
3. Double the result, and continue doubling the result until you can no longer keep track of the math in your head or until you feel warmed up, whichever comes first.

Depending on your memory, your skills at mental math, and how long you feel like warming up, you might want to repeat this process when you get stuck, picking a new small number each time.

The initial calculations might be trivially easy, but the difficulty increases as you go along, until you are juggling several numbers in your head and attempting to carry digits without dropping them. Don't worry about getting the answers right; the process is more important than the result. It will enhance the value of the exercise, however, if you are reasonably sure that your answers are correct. When you are unsure of a result, write it down and start the series again, double-checking your result when you get to the same place. The point here is that satisfaction is the stimulus that warms up your thinking.

Doubling numbers is essentially a mechanical process, so the mental challenge lies in remembering numbers as you apply the process. If doubling numbers bores you, you can try other forms of mental math, such as calculating second powers (i.e., *squares*). Memory games in general are a good source of mental warm-up exercises. If you are not naturally good with remembering names, for example, you can practice recalling names of casual acquaintances that you haven't seen in years. If you have used the Dominic System of mnemonics [Hack #6], you can practice it by picking a four-digit number and working out the mnemonic visualizations that go with that number and each successive number until you are warmed up.

Wordplay is another domain with mental-warm-up value. Try composing a limerick or a haiku to get your mental juices flowing. Puzzling out some rhymes or finding just the right words to fit syllable constraints can give you a small challenge and motivate your thinking. Another idea is to form anagrams using words you discover in your immediate environment.

How It Works

In his book *Flow*,[1] Mihaly Csikszentmihalyi describes the feeling of happiness you get when you engage in an activity that is challenging, but not beyond your abilities, and involves a continuous cycle of effort and success. The feeling of accomplishment motivates you to continue your efforts. This is true regardless of the level of challenge, as long as you do feel some sense of actual effort. Even a trivial success rewards your efforts and makes you more inclined to continue. As with the Swiss cheese technique [Hack #65], it is the small bursts of success that kick you out of your lethargy.

The distinction between a mental warm-up and hacks like the Swiss cheese technique is that a mental warm-up has no requirements. It is like stretching before you exercise. You can stretch at any time, and you need no equipment.

End Notes

1. Csikszentmihalyi, Mihaly. 1991. *Flow: The Psychology of Optimal Experience*. HarperPerennial.

—Karl Erickson

Play Board Games

Increase knowledge and hone a wide range of mental skills, painlessly and effectively, by exploring the fast-growing world of modern board gaming.

Maybe you think you know about board games, but if you're thinking of that Christmas present from your uncle with the pictures from your favorite cartoon on it that you had as a kid, stop right now. This is not your mom's Monopoly; board games are enjoying a worldwide renaissance in recent years, emerging in an ever-increasing array of sophistication, complexity, and ingenuity. In Germany, the nexus of modern "designer gaming," board gaming is so popular that a "game of the year" prize, the *Spiel des Jahres*, is awarded with great public interest. People everywhere and of all ages are getting hip to the fun, challenge, satisfaction, and intellectual development available for the taking by sitting down at a game table.

Modern board games are an excellent way to exercise and develop a variety of mental skills. In almost any game, you can improve your tactical and strategic ability, as well as learn the difference between tactics and strategy, and when to employ both. You can increase your understanding of statistics and probability, learn how to negotiate and make deals with other people, as well as play together toward a common goal, and become more adept at deduction and problem solving.

You can also learn raw information and gain a deeper, more integrated understanding of geography, sociology, military science, politics, and history. Board games can be the laboratory where you put theoretical mental skills into practice and strengthen them under testing and sparring situations. Board games are the dojo where your brain works out and goes from student to ninja. In short, you might get more out of a box of cardboard pieces and bits of wood and plastic than you might ever have dreamed.

In Action

To play board games, you obviously need other players and games. There are a lot of ways to find other players. You can start the old-fashioned way, by asking friends and family to play. See if any of them are interested in having a regular gaming session, once a month or more; it's a lot easier to sustain interest and build skills by working out regularly.

If you can't get your current friends to play games regularly, there are established gaming clubs and groups all over. If you live in a reasonably large metro area, you can almost certainly find more than one group to choose from, and even many smaller towns have a group or a game hangout. Try checking the bulletin board at your local game store, or asking the people who work there. If you live near a university, check the bulletin boards there or the student newspaper. You can also place ads or flyers yourself to attract other players; this is how the established groups get started, after all.

As it is with many other things, the Internet is also a terrific source of information about gaming groups. Use your favorite search engine to search the Web for your town and "board games", look under board games in directories like the Open Directory Project, or check blog and email group sites (such as Yahoo! Groups) for board-game players.

When you find players, you'll also need games to play. There are a dizzying number of games to choose from nowadays, so this can be daunting for someone trying to get a feel for what's available. Fortunately, when you find gamers, they'll usually have favorite games to suggest, so you can try those first to help narrow down your preferences. Nonetheless, once you get some idea of what you like and dislike, start branching out and trying new games on your own. You may well find a game you love that doesn't fit your other friends' tastes. You can't rely on other gamers, though, if you're starting up with your own friends and all of you are clueless.

Never fear, I'm here to help you out. Many board games fall into a few broad categories. Some fall into multiple categories at the same time. You'll probably want to try games from all of them to see what you enjoy:

Classic board games

Includes chess and its variants, Go, checkers (or draughts), Mancala, and Chinese chess. These games have been played for centuries, for good reason; you can't go wrong by making their acquaintance, and many people spend their lives trying to master them. They're also some of the most widely played games, so you can almost always find someone to play them, wherever you go. Because they're elegant, they're often simple to learn and will challenge basic skills such as strategy building, timing, assessing a situation accurately, and reading an opponent.

Modern classics

Includes Settlers of Catan, Puerto Rico, Acquire, Cosmic Encounter, Carcassonne, Magic: The Gathering, Tigris and Euphrates, and Reiner Knizia's Lord of the Rings. These are some of the most popular and well-respected board games in current play. You can usually find gamers who are happy to teach you, and you're likely to be rewarded by taking the time to learn them. These games come from many categories, so they cover lots of different skills, including tactical and strategic thinking, risk assessment, negotiation, and planning a sequence of actions.

Card games

Includes classic games such as bridge, hearts, pinochle, and poker, as well as new games such as Schotten-Totten, Frank's Zoo, Bohnanza, and Set. These games often improve your communication skills, deductive reasoning, and ability to work in partnerships, as well as strategic and tactical thinking.

Abstract strategy games

Includes games such as Chinese checkers, Blokus, Focus, Othello, Hex, and Twixt, as well as most of the aforementioned classic games. Abstract strategy games are stripped down to the very basics, removing themes and stories to leave stark and elegant mechanics that often turn on mathematical or geometric ideas. Abstract strategy games are good for developing abstract thinking, of course, as well as understanding geometric and math principles and enhancing problem-solving ability.

Party games

Includes games such as Taboo, Barbarossa, Pictionary, Cranium, and Balderdash. This is the kind of game that you'll probably encounter most often among friends who game casually. They tend to be less intellectually challenging, but they're better for building social skills, thinking quickly on your feet, and tuning creativity.

Strategic historical games

Includes games such as Age of Renaissance, Diplomacy, Advanced Civilization, Risk and its variants, Vinci, Parthenon, and History of the World. In this type of game, you act as leader of a culture or country (or perhaps several), and you must guide your country through historical pitfalls such as invaders, natural disasters, war, and international trade. This category also includes a whole subgenre of historical war games such as Battle Cry, which tend to focus on a specific time and place of military significance, such as the American Civil War. These games tend to be long and complex, but they can be very rewarding, especially in learning long-range strategy, negotiation, and a deeper understanding of the cultural and geographic underpinnings of history and politics.

Word games

Includes the classics Scrabble and Boggle, as well as newer games such as BuyWord and UpWords. Word games generally have to do with creating and placing words strategically, and they're excellent for increasing your linguistic understanding of how words are formed, as well as improving vocabulary and flexible thinking.

There are many, many other ways to categorize games, and this is just a start; I considered including sections about financial and auction games, race games, deduction games, and more, organized by game mechanics, themes, or subjective groupings. As you learn more games, you'll learn what you like and how other games are similar and different, and you'll have your own opinions about how to group them. There are also lots of wonderful games that don't fit into categories easily, but I hope this is enough of a map to get you started.

While you're exploring, there are a few things you should probably avoid, if you want to game with the goal of improving your brain. You probably won't get much out of "roll the die and move on the track" games; they generally have very simple mechanisms and high luck factors, and really won't challenge you much. These include the familiar games Sorry and Parcheesi, most popular themed games (such as games based on TV shows), and pretty much anything ending with "-opoly." You can learn basic facts and random information from trivia games, but most of them have a minimum of strategic potential and won't challenge those skills very much. Some games of chance are deep and can teach you a lot about strategy and deduction if you learn them seriously—poker is a good example—but many just revolve around the excitement of luck, and those won't build brain power either.

How It Works

Gaming, as an aspect of play, is rarely taken seriously as an important activity and a worthwhile way to spend time, but research in anthropology and psychology shows that play is an important method for learning and socialization throughout a person's life.[1,2] Why is it such a powerful way to learn?

First of all, gaming is fun. Rose and Nichols, among others, have shown that the brain is most efficient when it's enjoying an activity, and that activities that stimulate emotions as well as sensory input (such as shape, color, sound, and texture) are most effective at conveying information so that the brain retains it.[3] The colors, stories, laughter, and excitement involved in gaming open your mind to what it can learn as you play.

Gaming also provides a means for people to repeat situations safely and easily so that they can try out different decisions to see if they work. By studying academic courses that were based on games, Knotts and Keys determined that people who honed their skills in games, by trying out different decisions and getting immediate feedback about whether they were good decisions, developed more valid and sound skills than those who learned by more abstract methods.[4] By participating in the simulation-type situation that games provide, players also become actively involved with the information as they get into the game. This produces a "mindful" state in which people make more effort to think about what they're doing and not to rely on automatic reactions, which helps anchor information in their minds. Information also becomes easier to integrate because it's connected to a concrete situation and given a context, making it meaningful rather than inert.[5]

Playing games almost intrinsically encourages self-regulated learning, which is one of the most effective learning modes. Self-regulated learners are intrinsically motivated—they seek out learning experiences for their own value—and they participate actively in choosing and modifying the learning situation so that it best suits their needs.[6] Because games are fun, we want to play them, and the learning and intellectual challenge are built in as part of the fun. Games also provide a wide selection of attractive learning situations, and choosing them for stimulation as well as fun allows players to provide themselves with the best learning situation.

Much research about games and learning focuses on children, but one body of research related to self-regulated learning provides an important and relevant framework for adult learning and its motivations: the Flow Theory of Optimal Experience developed by Mihaly Csikszentmihalyi.[7] Flow theory gets its name from a particular state of extreme happiness and satisfaction produced by an activity. Flow occurs when people are so engaged and absorbed by an activity that they seem to "flow" along with it in a spontaneous and almost automatic manner, as if they are "carried by the flow" of the activity.

Flow comes from activities that provide enjoyment (as opposed to pleasure), which comes from one or more of the following factors:

- Challenge is optimized; the activity is neither too hard nor too easy.
- The activity absorbs all of the person's attention.
- The activity has clear goals.
- The activity provides clear and consistent feedback about reaching the goals.
- The activity is so absorbing that it frees the person from other worries and frustrations.
- The person feels completely in control of the activity.
- All feelings of self-consciousness disappear.
- Time is transformed subjectively during the activity, passing without notice.

This could be a verbatim list of reasons that dedicated gamers might give you if you asked them why they play. Achieving flow state takes a certain amount of effort and dedication, but its rewards in psychological growth and happiness are great, and they're only compounded by letting the pleasure of the flow state motivate you toward gaining the other intellectual benefits of gaming.

Finally, there is evidence that board games are among the mentally stimulating leisure activities that are associated with a lowered risk of dementia in aged people.[8] While the correlation doesn't prove a protective effect, gaming might have long-term benefits, which only adds a cherry on top of all the other positive effects.

In Real Life

As you might have guessed, I'm an avid board-game player myself. I generally play a minimum of one gaming session a week with my local group, and I often pick up a few games between meetings as well. I've played games since I was a child but have really only become deeply involved with them in the past five years or so. It has become such a rewarding and pleasurable experience for me, though, that I never intend to stop playing as much as I can.

One of my favorite games is *Age of Renaissance*. It's a long and complex board game, which follows a group of European cultures through several historical eras. Along the way, each develops technologies, endures calamities and bad luck, acquires trade markets and resources, battles and makes deals with the other cultures, and tries to accumulate the highest point score as measured by cultural and technological advancements.

This game challenges just about as many skills as I can bring to the table with me; I often lose, but it's so interesting and engaging that I'm still eager to play over and over. Playing *Age of Renaissance* has taught me how to build a long and complex strategy, how to use tactical moves to avoid sudden pitfalls and take advantage of good luck, how to negotiate wisely, how to assess probability and manage risk, and how to keep my focus on a goal instead of being distracted by side issues that might look interesting. I have taken every one of these skills and used it in other parts of my life, to my advantage.

Moreover, I have gotten an intuitive idea of how European culture developed and the cultural, economic, and geographical factors that influenced European history (and thereby the history of much of the world). I find now that when I am exposed to news about world politics, I have a very different view of its implications, which springs from my experiential knowledge of these historical simulations.

From other games, I've experienced more different simulated experiences than it would ever be possible for me to have in real life, and I've taken away something useful from almost all of them. I find new ways every day in which playing board games has improved my mind and changed my life for the better, and I can't wait to see what new adventures I'll have next week.

End Notes

1. Blanchard, K., and A. Cheska. 1985. *The Anthropology of Sport: An Introduction*. Bergin & Garvey Publisher, Inc.

2. Yawkey, T. D., and A. D. Pellegrini (Eds.). 1984. *Child's Play: Developmental and Applied*. Lawrence Erlbaum Associates.

3. Rose, Colin, and Malcolm J. Nicholl. 1999. *Accelerated Learning for the 21st Century*. Dell Publishing.

4. Knotts, S. Ulysses, Jr., and J. Bernard Keys. 1997. "Teaching Strategic Management with a Business Game." *Simulation & Gaming*, 28 (4): 337–393.

5. Salomon, G., D. N. Perkins, and T. Globerson. 1991. "Partners in cognition: Extending human intelligence with intelligent technologies." *Educational Researcher*, 20(3): 2–9.

6. Zimmerman, B. 1990. "Self-regulated learning and academic achievement: An overview." *Educational Psychologist*, 25(1): 3–17.

7. Csikszentmihalyi, M. 1990. *Flow: The Psychology of Optimal Experience*. Harper & Row.

8. Verghese, J., R. Lipton, M. Katz, C. Hall, C. Derby, G. Kuslansky, A. Ambrose, M. Sliwinski, and H. Buschke. 2003. "Leisure Activities and the Risk of Dementia in the Elderly." *New England Journal of Medicine*, 348: 2508–2516.

See Also

- BoardGameGeek (*http://www.boardgamegeek.com*) is a tremendous repository for all things related to board games. Look up games by name and see reviews and supplemental information, chat with other gamers on bulletin boards, see lists of recommended games, and get news about new games, events, and awards. An essential reference.

- Board Game Central (*http://boardgamecentral.com*) is another excellent source for board-game information. It's particularly notable because it contains several lists of game award winners, such as the *Spiel des Jahres* winners.

- Funagain Games (*http://www.funagain.com*) is one of the largest game retailers anywhere, and certainly the largest online. You can get almost any game you might want from the virtual shelves of Funagain and have it sent wherever you are. It's also an excellent information resource; it has listings for the *Games Magazine* Games 100 winners for several years. (Each year, *Games Magazine* selects the 100 best games of the year and awards winners in several categories; this is one way to find good games to play.)

—*Marty Hale-Evans*

Improve Visual Attention Through Video Games

#68 Some of the constraints on how fast we can task-switch or observe simultaneously aren't fixed. They can be trained by playing first-person-shooter video games.

Our visual processing abilities are by no means hardwired and fixed from birth. There are limits, but the brain's nothing if not plastic. With practice, we can improve the attentional mechanisms that sort and edit visual information. One activity that requires you to practice lots of the skills involved in visual attention is playing video games.

So, what is the effect of playing lots of video games? Shawn Green and Daphne Bavelier from the University of Rochester, New York, have researched precisely this question; their results were published in the paper "Action Video Game Modifies Visual Attention,"[1] available online at *http://www.bcs.rochester.edu/people/daphne/visual.html#video*.

Two of the effects they looked at are the *attentional blink* and *subitizing*. The attentional blink is that half-second recovery time required to spot a second target in a rapid-fire sequence. And subitizing is that alternative to counting for very low numbers (four and below), the almost instantaneous mechanism we have for telling how many items we can see. Training can both increase the subitization limit and shorten the attentional blink, meaning we're able to simultaneously spot more of what we want to spot and do it faster, too.

Shortening the Attentional Blink

Comparing the attentional blink of people who have played video games for four days a week over six months against that of people who have barely played games at all finds that the games players have a shorter attentional blink.

The attentional blink comes about in trying to spot important items in a fast-changing sequence of random items. Essentially, it's a recovery time. Let's pretend there's a video game in which, when someone pops up, you have to figure out whether they're a good guy or a bad guy and respond appropriately. Most of the characters that pop up are good guys, it's happening as fast as you can manage, and you're responding almost automatically—then suddenly a bad one comes up. From working automatically, suddenly you have to lift the bad guy to conscious awareness so that you can dispatch him. The attentional blink says that the action of raising to awareness creates a half-second gap during which you're less likely to notice another bad guy coming along.

Now obviously the attentional blink—this recovery time—is going to have an impact on your score if the second of two bad guys in quick succession is able to slip through your defenses and get a shot in. That's a great incentive to somehow shorten your recovery time and return from "shoot bad guy" mode to "monitor for bad guys" mode as soon as possible.

Raising the Cap on Subitizing

Subitizing—the measure of how many objects you can quantify without having to count them—is a good way of gauging the capacity of visual attention. Whereas counting requires looking at each item individually and checking it off, subitizing takes in all items simultaneously. It requires being able to give a number of objects attention at the same time, and it's not easy; that's why the maximum is usually about four, although the exact cap measured in any particular experiment varies slightly depending on the setup and experimenter.

Green and Bavelier found the average maximum number of items their non-game-playing subjects could subitize before they had to start counting was 3.3. The number was significantly higher for games players: an average of 4.9—nearly 50% more.

Again, you can see the benefits of having a greater capacity for visual attention if you're playing fast-moving video games. You need to be able to keep on top of whatever's happening on the screen, even when (especially when) your attention is stretching.

How It Works

Given these differences in certain mental abilities between gamers and non-gamers, we might suspect the involvement of other factors. Perhaps gamers are just people who have naturally higher attention capacities (not attention as in concentration, remember, but the ability to keep track of a larger number of objects on the screen) and have gravitated toward video games.

No, this isn't the case. Green and Bavelier's final experiment was to take two groups of people and have them play video games for an hour each day for 10 days.

The group that played the classic puzzle game *Tetris* had no improvement on subitizing and no shortened attentional blink. Despite the rapid motor control required and the spatial awareness implicit in *Tetris*, playing the game didn't result in any improvement.

On the other hand, the group that played *Medal of Honor: Allied Assault* (Electronic Arts, 2002), an intense first-person shooter, could subitize to a higher number and recovered from the attentional blink faster. They had trained and improved both their visual attention capacity and their processing time in only 10 days.

In Real Life

Green and Bavelier's results are significant because we continuously use processes like subitizing in the way we perceive the world. Even before perception reaches conscious attention, our attention is flickering about the world around us, assimilating information. It's mundane, but when you look to see how many potatoes are in the cupboard, you'll "just know" if the quantity fits under your subitization limit, and you'll have to count them—using conscious awareness—if it doesn't.

Consider the attentional blink, which is usually half a second (for the elderly, this can double). A lot can happen in that time, especially in this information-dense world: are we missing a friend walking by on the street, or cars on the road? These are the continuous perceptions we have of the world, perceptions that guide our actions. And the limits on these widely used abilities aren't locked but are trainable by doing tasks that stretch those abilities: fast-paced computer games.

I'm reminded of Douglas Engelbart's classic paper "Augmenting Human Intellect,"[2] on his belief in the power of computers. He wrote this in 1962, way before the PC, and argued that it's better to improve and facilitate the tiny things we do every day than it is to attempt to replace entire human jobs with monolithic machines. A novel-writing machine, if one were invented, just automates the process of writing novels, and it's limited to novels. But making a small improvement to a pencil, for example, has a broad impact: any task that involves pencils is improved, whether it's writing novels, newspapers, or sticky notes. The broad improvement brought about by this hypothetical better pencil is in our basic capabilities, not just in writing novels. Engelbart's efforts were true to this: the computer mouse (his invention) heightened our capability to work with computers in a small, but pervasive, fashion.

Subitizing is like a pencil of conscious experience. Subitizing isn't just responsible for our ability at a single task (like novel writing); it's involved in our capabilities across the board, whenever we have to apply visual attention to more than a single item simultaneously. That we can improve such a fundamental capability, even just a little, is significant, especially since the way we make that improvement is by playing first-person-shooter video games. Building a better pencil is a big deal.

End Notes

1. Green, C. S., and D. Bavelier. 2003. "Action video game modifies visual attention." *Nature*, 423: 534–537.

2. Engelbart, D. 1962. "Augmenting Human Intellect: A Conceptual Framework." *http://www.bootstrap.org/augdocs/friedewald030402/augmenting humanintellect/ahi62index.html.*

—Matt Webb and Tom Stafford

HACK #69 Don't Neglect the Obvious: Sleep, Nutrition, and Exercise

You know as well as we do that you need to sleep well, eat right, and exercise for your brain to be in peak condition. We're not going to scold you, but we are going to present some information you might not have known about your brain's relationship to your body.

Many tips for mental performance concentrate on extending productivity for a short while after tiredness kicks in, or aiming for a slightly higher *peak* during a work period. Importantly, all of these increases in performance are relative to your *normal* or baseline level of functioning. You can often obtain more substantial and more widely effective gains in mental ability by making sure that your baseline performance is at an optimum. Tuning sleep, nutrition, and exercise is one effective way of doing this.

The brain, like any other organ in the body, works best when it is optimally fueled and is given adequate time to recover after periods of extended exertion or effort. Here, *fuel* does not mean energy in the form of only those foods that are broken down into glucose, but also those that provide essential nutrients needed for a wide range of complex functions. Neuroscience research has now identified a number of brain nutrients that can result in varying degrees of mental impairment if a deficiency exists.

On the other hand, sleep is still a bit of a mystery. Despite the fact that it takes up about a third of our time, we know surprisingly little about why we sleep. The nearest to a current consensus among scientists is that sleep, and particularly REM sleep, makes sense of disparate, emotionally fragmented, or weakly coupled memories and weaves them into a coherent structure that the brain can use more effectively during wakefulness. It is not clear, however, whether this theory is popular because memory is easy to test and so provides plenty of supporting evidence, or whether the function of sleep might be much broader, but evidence for the other functions has thus far been harder to come by. Either way, it is clear that lack of sleep causes a whole range of cognitive problems, suggesting that it fulfills an important role in maintaining the mind and brain at their peak.

Exercise is known to have beneficial effects on mental performance, both for the short-term oxygen boost it provides to the brain and for the important role that it plays in maintaining a healthy and efficient blood supply to the brain. The system of arteries and veins is important for providing essential nutrients and for removing dangerous toxins, and is so highly tuned that it can adjust in less than a second to take account of changing mental demands. This, however, requires an efficient, smoothly operating transport system (known as the *cerebrovascular system*), which is maintained and improved by regular exercise.

In Action

You can take some simple and practical steps with respect to sleep, nutrition, and exercise to improve your mental performance.

Sleep. A lack of sleep produces some of the most striking impairments in mental performance, as anyone who has pulled an all-nighter will know. A good night's sleep [Hack #70] is particularly important for memory[1], and there is now increasing evidence that things learned shortly before sleep are remembered better than those learned earlier in the day. It also seems that as something becomes more complex, the role of sleep is more important in efficiently remembering it. This applies equally to remembering skills involving body movements (for example, learning to juggle) and to remembering verbal or purely mental information.

The corollary of this—that sleep deprivation can negatively affect mental performance and muscle control—has been borne out by a number of studies that have shown that sleep deprivation can also result in mood disturbances.[2] This suggests that skimping on sleep to give more time to learn can be counterproductive, as each hour spent sleep deprived is worth only a part of an hour fully rested.

One of the more surprising findings is that lack of sleep does not affect mental function only, but is related to a decrease in almost all measures of long-term health, including risk of heart disease, diabetes, high blood pressure, and inflammation, to name but a few.[3] These studies have led doctors to suggest that sleep should not be considered a luxury, but an important component of a healthy lifestyle, and therefore essential as a factor in maintaining cutting-edge brain function.

Nutrition. Following an all-around healthy diet ensures that an organ as sensitive as the brain has all the necessary resources available to work at an optimal level. Some nutrients and diet options have been particularly linked to maintaining a sharp mind. Some of the more unusual ones [Hack #73] are discussed in this book. Here, however, are some of the more well-known nutrients, although not everyone is aware of their importance.

Vitamin B12 and folic acid (also known as *folate*) are known to be important in mental performance.[4] Adequate levels of these nutrients are vital, because they play a role in the functioning of the nervous system, including the creation of neurotransmitters (specifically dopamine, epinephrine, and norepinephrine), as well as keeping levels of a risky amino acid, called *homocysteine*, to a minimum. High levels of homocysteine are now thought to be a major risk factor for poor health, with consequences including impaired brain function and possible damage to the heart, which can lead to a double whammy if the brain's blood supply is affected.

Various fats are now known to affect how the brain works. A diet low in saturated fat and high in cereal and vegetable foods is known to promote good cognitive function, as is a diet high in certain omega-3 fatty acids, which are present in flax seed, walnuts, and oily fish.[5] These fatty acids are now thought to be so important that omega-3 supplements are now being tested as effective ways of improving mood and cognition in certain types of mental illness.

Antioxidants prevent the oxidation of other chemicals, a process known to produce tissue-damaging substances called *free radicals*. Several essential nutrients are antioxidants, including vitamins A, C, and E, and the trace element selenium (present in Brazil nuts). Low levels of antioxidants have been linked to an increased risk for a number of brain disorders, including Alzheimer's disease, but it is not clear whether there is a link to mental function in healthy young people. It does seem, however, that a good intake of these nutrients may protect against cognitive decline in later life.[6]

Breakfast is often called the most important meal of the day, and there is some evidence to support this claim. Missing breakfast has been consistently linked to poor mental performance, particularly on memory and visual recognition tasks,[7] at least in children, on whom most of this research has been carried out. There is some evidence that high-fiber foods that release energy slowly (such as some cereals) and high-protein foods may be particularly good for keeping your edge throughout the morning, and low-fiber, high-carbohydrate foods (such as pastries) might cause a period of slower, fuzzier thinking.

Exercise. Walking for 30 minutes a day five days a week is typically recommended as adequate exercise for significantly improving all-around health. It has the medium- to long-term effect of strengthening the heart, reducing blood pressure, and even lifting mood. All of this is good news for sharpening the mind, which does better with a healthy brain and positive outlook. In the short term, any activity that boosts oxygen intake will immediately affect mental performance for the better,[8] so as long as it is not too distracting, any light exercise should help you learn while you take part.

In Real Life

The link between brain function and sleep, nutrition, and exercise is still only partially understood, so the recommendations for healthy daily amounts change over time as new research emerges. Keep up with the latest in health advice to make sure you can tune your life for optimal brain function.

One consistent finding is that middle-aged and older adults tend to show a greater detriment in mental function due to poor diet, sleep patterns, and exercise than younger adults do. If you are middle-aged or older, paying particular attention to these health issues will ensure you keep your edge well into old age. If not, there's no time like the present to establish healthy habits that will bring you benefits for the rest of your life.

End Notes

1. Stickgold, R., J.A. Hobson, R. Fosse, and M. Fosse. 2001. "Sleep, learning, and dreams: off-line memory reprocessing." *Science*, 294: 1052–1057.

2. Durmer, J.S., and D.F. Dinges. 2005. "Neurocognitive consequences of sleep deprivation." *Seminars in Neurology*, 25: 117–129.

3. Alvarez, G.G., and N.T. Ayas. 2004. "The impact of daily sleep duration on health: a review of the literature." *Progress in Cardiovascular Nursing*, 19: 56–59.

4. Stabler, S. 2003. "Vitamins, homocysteine, and cognition." *American Journal of Clinical Nutrition*, 78: 359–360. *http://www.ajcn.org/cgi/content/full/78/3/359*.

5. Kalmijn, S., M.P. van Boxtel, M. Ocke, W.M. Verschuren, D. Kromhout, and L.J. Launer. 2004. "Dietary intake of fatty acids and fish in relation to cognitive performance at middle age." *Neurology*, 62: 275–280.

6. Gray, S.L., J.T. Hanlon, L.R. Landerman, M. Artz, K.E. Schmader, and G.G. Fillenbaum. 2003. "Is antioxidant use protective of cognitive function in the community-dwelling elderly?" *The American Journal of Geriatric Pharmacotherapy*, 1: 3–10.

7. Rampersaud, G.C., M.A. Pereira, B.L. Girard, J. Adams, and J.D. Metzl. 2005. "Breakfast habits, nutritional status, body weight, and academic performance in children and adolescents." *Journal of the American Dietetic Association*, 105: 743–760.

8. Scholey, A. 2001. "Fuel for thought." *The Psychologist*, 14: 196–201. *http://www.bps.org.uk/_publicationfiles/thepsychologist%5CScholey.pdf.*

See Also

- BBC Health information; *http://www.bbc.co.uk/health.*

—Vaughan Bell

HACK #70   Get a Good Night's Sleep

By programming the associations your brain makes as you start to feel sleepy, you can set yourself up for a good night's sleep—or an awful one.

Here's the straight dope about getting a good night's sleep. There's no miracle method here, no secret that will let you survive and thrive on less than four hours a night. But hopefully, you should be able to make sure that you spend more of the time you allocate to sleeping actually asleep, and spend less of the time you want to be awake sleepy.

Everybody has slightly different preferences about when and how they sleep. The idea of morning people and evening people is solid fact, not myth.[1] People also vary in how much sleep they need, and how they feel when they don't get it. We don't need to be macho about it. If you need nine hours a night, that's normal; don't starve yourself. Many of the people who supposedly need very little sleep are nappers, such as Winston Churchill and Margaret Thatcher, two British prime ministers who famously got by on four hours a night or less. Others kept up sparse sleep habits for short periods only when they had big projects going, such as Leonardo da Vinci, who reported that he'd sleep for 15 minutes every four hours.[2]

Too little sleep is bad for your health and it will make you feel rotten, even if you are awake. That said, too much sleep is bad for you as well.[3] The average is seven to eight hours a night, with variation among individuals about how much and exactly when they want to take it. This hack is about quality and efficiency of sleep, about making sure that when you want to be asleep you are, and that when you do sleep it's good sleep.

Sleep is nourishment for your mental life. Although my guess is that we've evolved to want more sleep than we really need (as with food), and we all can certainly get by for a few nights with very little sleep, trying to establish a long-term routine that deprives you of sleep will harm your mood, your memory, and your performance during the day. Plus, you'll be missing out on one of life's great pleasures. So, beyond a wholesome discipline, I recommend being gentle on yourself.

Sleep Hygiene

Really, this is a story about associations. Your brain constantly scans the environment and puts things together in memory. Mental connections are made between things that occur at the same time, or in the same place, or that are followed by good or bad consequences. Continuously, often unconsciously, your brain is associating.[4] We can use this to help get a better night's sleep and to understand some of the bad sleep habits that you might have fallen into.

The key idea here is *sleep hygiene*, which means keeping those things you associate with sleep totally separate from those things you associate with being awake. We'll assume that, being a sensible person, you already do all the obvious things that will help you sleep. As a reminder, some of those are:

Making a comfortable bed
Create for yourself whatever a comfortable bed means to you, hard or soft, big or small, in whatever location you find most relaxing.

Being able to keep warm or cool all night
Provide yourself with your favorite blankets, fans, or whatever you need.

Being somewhere quiet
If your room isn't quiet enough, lots of people swear by earplugs or white-noise machines.

Being somewhere dark
Likewise, lots of people swear by eye masks and blackout curtains when they can't find a dark room.

Being well nourished, but not eating heavily immediately before sleep
Especially be sure to get enough iron, lack of which can cause daytime fatigue.

Cutting out caffeine at least eight hours before bed, and not drinking alcohol
Both of these drugs will reduce the quality of sleep you get, even if they don't reduce the quantity.

Making Associations

So, if your sleep isn't being artificially affected, and the physical conditions are ripe for a good night's sleep, the best way to ensure that you get it is to strongly associate when and where you sleep with sleep, and sleep only. You train the associations by your behavior, so your brain picks up the regularities in your experience and ingrains them as habit. Three kinds of associations to think about are those relating to *object and location*, *time*, and *behavior*.

Object and location. Associations in object and location means, simply, that you should use your bed, and the other things around you when you sleep (e.g., nightwear), purely for sleep and nothing else.[5] If you wander around in your pajamas in the morning for hours, or sit in bed and read till the small hours (or worse, use a laptop), you're sending a confusing signal to your brain. "Is this sleep time?" your brain will be asking, "or is it awake time?"

When you get into bed, go to sleep; when you wake up, get out of bed immediately and get changed. If you can't sleep for 20 minutes, get up and be awake somewhere else. You need to preserve the association of your bed with being asleep, not with trying to get to sleep and worrying about it. This is why study-bedrooms are such a disaster. When you are working, you look at your bed and feel sleepy. When you are trying to sleep, you look at your work and think about work.

Time. Setting aside dedicated space is a good method for any activity that requires a certain frame of mind, not just sleep. Even better is if you can set aside dedicated space and regular dedicated time. Regular time is particularly important if you can't schedule dedicated space, and vice versa.

A regular bedtime works well for children, but most of us don't have the lifestyle that permits going to bed at the same time every night, even if we know that this habit would help us sleep better. One thing that works well for many people is to have a fixed wake-up time and being flexible with going to bed.[6] This gives you the routine that will train your body to wake up and be awake in the morning, but also the flexibility that lets you go to bed when you are actually tired, rather than when you think you should be.

After a while of doing this, your body will learn to wake you up in the morning, and because your body clock is expecting that fixed waking-up time, it will learn to make you feel tired at an appropriate time the evening before (with accommodation for seasonal variations in sleep requirements, and other personal fluctuations that happen more or less at random). And if you ever do need to stay up late (work? babies? friends who are incorrigible evening people?), you are not breaking the association of your wake-up time with being awake, and the routine you've established should carry you through any extra bit of tiredness you incur.

Behavior. The third kind of associative purity is associations between what you *do* and sleep. Although some of us don't need any cues to just drop off, most of us need to wind down a bit before we slip away into the land of dreams. If you establish a fixed ritual before you go to bed, it will tell all the unconscious and semiconscious parts of your mind clearly that *now it is time to sleep*. It doesn't have to be a long list; maybe just perform your ablutions, change into whatever you wear at night, turn off the light, and then hit the pillow. The important thing is that whatever you do, you don't do any of these things long before you go to bed, or do them out of order, or mix in anything not sleep related. For example, getting into your pajamas and then doing 30 minutes on the StairMaster is a recipe for troubled sleep in the future.

Likewise, when your alarm goes off in the morning, get up immediately and change clothes. If you're having particular trouble training yourself in a getting-up habit, you might try incorporating some morning physical exercise, a particularly un-sleeplike association to develop. If you have trouble getting to sleep, you might want to put something in your pre-bed routine to make it longer, preferably something you can do until you are properly ready to go to bed. Something like reading is ideal, because it is gentle and it is an enjoyable way to wait until you are sleepy.

Structure versus flexibility. Setting up and maintaining pure associations between sleeping time and awake time should help you be asleep when you want, and wake up when you want as well. The cost is a loss of flexibility, but the benefit is spending less time trying to sleep and not being able to.

The structure provided by routine should carry you through times when tiredness might otherwise make you fall asleep. You might decide that you value the flexibility of being able to break all these associations more than the benefits of creating and maintaining them. This is great, if it works for you. And there's some good evidence that when it comes to sleep, whatever you do, it is vital not to take things too seriously and that worrying may be the worst thing of all.

Hacking the Hack

Our perception of time is radically distorted when we are asleep or close to sleep.[7] By attaching electronic monitors to insomniacs, it is possible to show that many who report sleeping only two or three hours a night are actually asleep for an average of seven hours—only 35 minutes less than the average for non-insomniacs! So, for insomniacs, or the rest of us on a bad night, when we feel like we "hardly slept at all," we were probably asleep for most of the night but only remember vividly, and exaggerate, the time when we were trying to get to sleep. Not only this, but if you tell insomniacs that they actually slept better than they did, they then feel and perform better during the day.[8]

Other results show that it is impossible to prove an extra effect of broken sleep beyond the amount of sleep time you lose from being awake.[9] In other words, if a screaming baby wakes you up in the middle of the night for 20 minutes, there's no reason that this should make you feel any worse in the morning than going to bed 20 minutes later than normal would have. Any discontent you do feel is probably due to stress, rather than lack of sleep per se.

This means that it's how you feel about how much sleep you got, as well as how much sleep you actually got, that matters. If you can avoid worrying about it, avoid watching the clock and counting the hours, that's as useful as actually getting a few more minutes. Obviously, most insomniacs don't *want* to worry about not getting enough sleep, so does knowing this help? Well, hopefully it helps a little bit. When you think you haven't slept enough, you probably have. As long as you cover a core minimum of sleep, which is probably a couple of hours less than you normally get, you might feel tired but your performance won't suffer. If you know that awakening in the middle doesn't have to matter that much, you are free not to worry excessively about how much sleep you get. And the beautiful thing is that if you don't worry about it, you'll sleep better and you truly will require less sleep overall.

End Notes

1. This post is a highly recommended place to start finding out about the science of sleep, including the details of larks and owls: "Everything You Always Wanted To Know About Sleep (But Were Too Afraid To Ask)." *http://circadiana.blogspot.com/2005/01/everything-you-always-wanted-to-know.html.*

2. This kind of thing isn't sustainable over the long term, and would be a nightmare for everyone you knew anyway.

3. Kripke, D., et al. 2002. *Mortality Associated With Sleep Duration and Insomnia.* The Archives of General Psychiatry, 59: 131–136. Also, see "Too much sleep 'is bad for you'," at *http://news.bbc.co.uk/1/hi/health/ 1820996.stm*, which notes, "A study that included more than a million participants found people who sleep eight hours or more died younger." (Although this is just an average, what counts as *too much* or *too little* depends on your individual requirement.)

4. This power of association is called *conditioning* in psychology. I wrote a bit more about it in *Mind Hacks*, where I talked about how caffeine, because it is a drug of reward, can rewire our brains to make us obsessive about how our coffee is prepared and served. A PDF of the draft hack is available from the O'Reilly page for the book, at *http://www.oreilly.com/catalog/mindhks/chapter/hack92.pdf*.

5. Well, maybe for *one* other thing!

6. See Steve Pavlina's "How to Become an Early Riser" at *http://www.stevepavlina.com/blog/2005/05/how-to-become-an-early-riser* for a great piece discussing how the author turned himself from an evening person to a successful morning person using the fixed-rising-time method.

7. Semler, C.N., and A.G. Harvey. 2005. "Misperception of sleep can adversely affect daytime functioning in insomnia." *Behaviour Research and Therapy*, 43: 843–856. Also, see "You're feeling very sleepy" at *http://bps-research-digest.blogspot.com/2005/05/youre-feeling-very-sleepy.html*.

8. Tang, N.K.Y., and A.G. Harvey. 2004. "Correcting distorted perception of sleep in insomnia: a novel behavioural experiment?" *Behaviour Research and Therapy*, 42. 27–39. Also, see "Don't think, sleep!" at *http://www.mindhacks.com/blog/2004/12/dont_think_sleep.html*.

9. Wesensten, N.J., et al. 1999. "Does sleep fragmentation impact recuperation? A review and reanalysis." *Journal of Sleep Research*, 8: 237–245.

—*Tom Stafford*

Navigate Around the Post-Lunch Dip
#71 Recognize the differences in your state of mind at different times of the day.

One function of your *biological clock* is to adjust your level of wakefulness during the day. As a general rule, concentration and logical reasoning peak between 10 a.m. and 2 p.m., and alertness peaks between 4 p.m. and 8 p.m. A sleepy feeling in the early afternoon known as the *post-lunch dip* is common. A large lunch with alcoholic drinks will clearly contribute to the effect, but the effect is present even without it.

These patterns aren't the only diurnal fluctuations in awareness, which vary from person to person. Mapping your own variation in *mindset* over the course of the day will allow you to plan accordingly. The intention is to use the pattern to your advantage, instead of trying to maintain some optimal state.

In Action

People vary a great deal in their *chronotype*: the way specific times of day affect them. As a rule, older people tend to be morning types and younger people more alert in the early evening, but this is only a general trend. However, the post-lunch dip, a low point of wakefulness in the circadian rhythm that coincides with the process of body chemistry switching to digestion after a meal, is something many people experience.

Here are a few suggestions for dealing with the post-lunch dip:

- Experiment with changes in what you eat for lunch. Try reducing the amount of carbohydrate-rich food, as well as increasing the variety and reducing the quantity of your food. Although my colleagues find it a bit odd, I find mugs of hot water through the day help me stay sharper, especially for the hour or so after lunch. Tastes and food sensitivities vary, so try making changes and see what works for you. One tip: the duration of effects of food varies, and the intensity of these general tendencies will vary from person to person. Sugars (in large quantity) have rapid onset, but their effects tail off quite quickly. Hydrogenated oils have a slower onset, and their influence can last for hours.

 You can use the rapid onset of sugar's effects **[Hack #72]** to enhance learning.

- Have lunch right at the start of your lunch break, and then spend the remainder of the break taking a walk. Less of your post-lunch dip happens while you're trying to work after lunch, and the exercise helps wake you up and walk off stress. It also speeds the digestion process and can improve your body's uptake of glucose from the blood, which may shorten the *rebound* time after eating.

- Lastly, adjust the timing of specific tasks so that you work *with* your mindset. Choose tasks for after lunch that need to be done, but that don't require especially intense thought. Active *tidying-up* tasks are a good choice.

How It Works

Taking time to learn how the mind changes during the day allows one to plan for it. The simple act of taking a walk at lunchtime can have much more benefit to thought than it might seem. It is quite effective in reducing work-related stress. Unloading such stress has a positive effect on clarity of thought. The post-lunch dip is an almost ideal time for unwinding. It's a time of naturally reduced alertness, so it becomes easier to shake off stress than when the body is *on alert*.

Scheduling tidying-up tasks for after lunch works for similar reasons. Because the tasks aren't highly demanding of attention, the mind can tidy itself in the background at the same time, making it easier for the tidied mind to be more efficient an hour later.

In Real Life

The variation in mindset within a day also suggests why writers should find a writing routine that suits them, a standard time of day during which to write. I've been part of an amateur fiction writers' group for many years. One problem I've often had was picking up a story I'd partly written earlier and continuing it in another session. I'd find the pieces "didn't join up" at the point where I'd put down the writing and then picked it up again. I was writing in a different mood when I started again.

A factor in this became clearer when, for a couple of months, I started waking early and writing for an hour each morning. My hope was that by writing just after waking, I would be more creative than normal. I found that when I wrote this way I had no problem with the pieces joining up. Unfortunately, at that time in the morning, my style of writing was too surreal and whimsical. It was consistent, but not in a way that I wanted. Although the experiment didn't work the way I had hoped, it showed me that the principle is valid: mindset changes during the day.

See Also

- Foster, R., and L. Kreitzman. 2004. *Rhythms of Life*. Profile Books. *Rhythms of Life* describes alertness and other temporal variations and examines the biological basis for the circadian rhythm.

—James Crook

HACK #72 Overclock Your Brain

In some situations, the brain is performance-limited by the available fuel. Increase the fuel and you can temporarily get a performance boost.

The brain is one of the most energy-hungry of the human organs. Despite making up only about 2% of the average body weight, it uses almost 20% of the normal intake of energy. Although the brain comprises mostly fat, this is mainly used to protect and insulate brain cells and is not available as an energy store. It therefore relies on the rest of the body to provide it with a supply of energy, which consists almost entirely of glucose. The brain uses up its own glucose supplies in about 5–10 minutes if they are not replenished, meaning it is particularly sensitive to changes in blood glucose levels.

As part of this process, oxygen is also needed and is another essential component of the brain's fuel supply. Oxygen is used as part of glucose metabolism to provide brain cells with a number of important chemicals that allow them to support themselves and communicate with other neurons.

Mental performance relies on the functioning of the brain, and like with any other organ, this performance is linked to how many resources are available. Research has shown that in some instances, mental performance is rate-limited by the available glucose and oxygen. In other words, you can increase the rate of mental processing by increasing the available fuel.

It turns out that this effect is not global, and it typically affects some mental abilities more than others. To get the best performance increase, you need to know how quickly glucose and oxygen are metabolized in the body to perfect your timing, and which mental processes are most affected to select your task.

In Action

One of the most reliable findings is that increasing available glucose and oxygen seems to have a beneficial effect on memory. Importantly, the effect is usually found for memory encoding but not memory recall. If you are not familiar with this distinction, think of it in terms of the mental activities involved in memory. Encoding is when you encounter the information and try to commit it to memory, and recall is when you want to retrieve previously committed information.

Increasing glucose and oxygen supplies to the brain seems to allow information to be committed more accurately and fully to memory; in other words, you learn better. This means when you come to recall it at a later stage, you will undoubtedly do better, because the information there is clearer and more comprehensive. The reverse does not seem to be true, however. If you first encoded something without the aid of extra oxygen and glucose, suddenly making more oxygen and glucose available when you try to recall it will not improve your overall memory performance.

Boost oxygen levels. The improvement in oxygen levels on memory[1] typically lasts for a few minutes only (five is about the limit), so you need to time your learning to happen shortly after an increase in oxygen, or ensure that you maintain a slightly increased level for the duration of the learning period. Oxygen canisters are available in some shops, although they are often expensive and unwieldy. More usefully, deliberately taking some deep breaths will increase blood oxygen levels for a short time, as will light exercise. Going for a walk while listening to something you want to remember on an MP3 player should do the trick, as long as the environment is not so distracting that you cannot concentrate.

Optimize glucose supplies. Glucose has a much longer-term effect, as shown in Figure 8-1.[2] Here the maximum available glucose peaks at about an hour, although it rapidly becomes available after it has been ingested. All energy-giving foods are broken down into glucose at some stage, although at different rates. This graph charts the rate of pure glucose absorption, so it best matches the effects of sugary drinks.

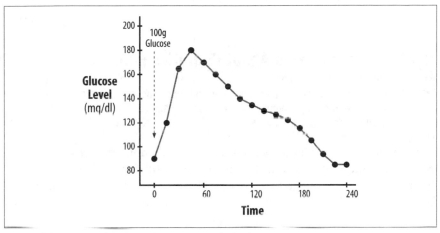

Figure 8-1. Graph of glucose absorption in the body

Glucose is important as a simple fuel, but it is also used in the creation of the neurotransmitter acetylcholine. This brain chemical is particularly linked to memory, and it's no accident that, like oxygen, extra glucose is linked to an increase in memory and learning ability.[3] Again, timing is crucial, but not so much effort is needed to constantly maintain glucose levels. A well-timed sugary drink, 30 minutes to an hour before you want to remember or take notice of something particularly closely, should improve how well you remember it.

End Notes

1. Scholey, A.B., M.C. Moss, and K. Wesnes. 1998. "Oxygen and cognitive performance: the temporal relationship between hyperoxia and enhanced memory." *Psychopharmacology*, 140: 123–126.

2. Messier, C. 2004. "Glucose improvement of memory: a review." *European Journal of Pharmacology*, 490 (1-3): 33–57. Figure 8-1 is my own simplified re-creation of a graph from this article.

3. Meikle, A., L.M. Riby, and B. Stollery. 2004. "The impact of glucose ingestion and gluco-regulatory control on cognitive performance: a comparison of younger and middle-aged adults." *Human Psychopharmacology*, 19 (8): 523–535.

See Also

- "Fuel for thought" by Andrew Scholey. *http://www.bps.org.uk/_ publicationfiles/thepsychologist%5CScholey.pdf*.

—Vaughan Bell

Learn the Facts About Cognitive Enhancers

#73 "Smart drugs" are supposed to make you smart, but it's not always smart to take them. Until smart drugs are safer and more effective, there are some alternative mental-performance enhancers to try that are both interesting and legal.

Mind-altering substances have been used for millennia to alter how we perceive and understand the world. Some of these substances have been taken because they are thought to enhance specific aspects of our thought and behavior to enable us to become more productive—caffeine is a popular example.

More recently, advances in drug testing and development and a better understanding of how the brain works have resulted in drugs that are intended to boost intelligence or cognition in specific ways. Often, these drugs have been developed to help with specific illnesses or conditions, but they are now gaining notoriety, as they are being used illegally by people hoping to improve performance during intense work or study periods. Others are still in development and are currently only in the experimental stages:

Pharmaceutical amphetamines (such as Adderall and Ritalin)
These drugs have been used without a prescription, for their tiredness-reducing and concentration-enhancing effects, by about 5% of students in U.S. colleges, according to a recent study, with some colleges reporting rates as high as 25%.[1]

Modafinil
Modafinil is a nonamphetamine stimulant, marketed to help people with narcolepsy stay awake. It has been reported as popular with otherwise healthy individuals who are using it to maintain concentration levels and alertness during several days of wakefulness.[2]

Ampakines

Ampakines are a class of drugs known to affect the sensitivity of the brain to a neurotransmitter called glutamate, a chemical known to be important in fast information transfer and memory formation. These drugs are still in development, but one, known as CX516 and currently targeted at treating schizophrenia, has been shown to improve cognitive performance in the elderly[3], and other ampakines, such as CX717, are being touted as future cognition-enhancing drugs.

MEM 1003

MEM 1003 is a substance being developed by a company advised by the Nobel Prize–winning neuroscientist Eric Kandel. Although few details are available, it is to be marketed for clinically diagnosable conditions, but also for "age-associated cognitive decline" in healthy individuals. In other words, it's a chemical pick-me-up for those experiencing the normal decline in memory that typically occurs past middle age. The fact that this drug is being openly promoted for use in people without serious illness suggests that cognitive enhancers may become increasingly mainstream.

Most of these drugs are still in the pipeline, and for the time being most "cognitive enhancers" are officially classed as medicines. Nonmedical uses of these substances are officially frowned upon, and possession could lead to jail time in some countries. For some people, however, a greater risk may be unwanted short- or long-term effects of such drugs. It is well known that amphetamines can trigger psychosis in some users, particularly with heavy use, and the fact that most people will use them without proper medical advice and assessment makes it unlikely that any negative effects will be detected sufficiently early. Newer drugs are typically marketed as being free of side effects, although history tells us that some side effects do not come to light until later on.

For those wanting to avoid pharmaceuticals, the scientific literature has reported some alternative ways of temporarily boosting mental performance, some more unusual than others.

In Action

Some foods and over-the-counter drugs may significantly enhance your mental performance, if taken in moderation.

Glucose. Glucose is the brain's fuel, and there is plenty of evidence that a well-timed glucose intake increases mental performance and that you can use sugary drinks to overclock your brain [Hack #72]. Glucose is broken down to become an important element in brain-cell function, as well as being important in the creation of a neurotransmitter called acetylcholine. This chemical is particularly linked to memory, and it is no accident that extra glucose is linked to an increase in memory and learning ability.[4] In the long term, too much sugar can lead to health problems, but in the short term it can give a brief mental lift.

Ginkgo biloba. Gingko biloba is an ancient plant that has survived many thousands of years and has no living relatives. It's commonly sold in health-food stores as an herbal supplement, and there is good evidence that in small doses it can increase mental performance, particularly attention, in healthy young adults.[5] It is thought to work through direct effects on neurotransmitters and by promoting blood flow and circulation to the brain. Although ginkgo is considered safe enough to be sold in shops, it does not agree with everyone. Some people may find it upsets their stomach. More seriously, it can act as a blood thinner, and it's usually advised that people with blood circulation disorders, those taking aspirin, pregnant women, and people taking certain forms of antidepressants (known as monoamine oxidase inhibitors or MAOIs) should avoid it as a precaution. If in doubt, discuss it with your doctor.

Chewing gum. Chewing gum? A number of recent studies have found that chewing gum improves mental performance, typically memory.[6] It's still not clear exactly why this happens. Speculations include the fact that chewing causes insulin to be released in the body in anticipation of food digestion, which increases the rate of glucose uptake. This might make more glucose available for the brain, temporarily providing more fuel for cognitive functions. Another theory is simply that chewing increases arousal, making us slightly more alert and therefore that little bit sharper.

Tyrosine. Tyrosine, an amino acid, is one of the building blocks of a group of neurotransmitters called the catecholamines, which include dopamine, epinephrine, and norepinephrine. It's commonly available as a supplement in health-food shops and has been shown to significantly increase mental performance, particularly during periods of tiredness.[7] In some people, however, it can trigger migraines and stomach upset, and people taking antidepressants and stimulants are usually advised to avoid it to prevent potentially harmful interactions.

Ginseng. Ginseng is a plant extract that is taken throughout the world, and a number of claims have been made for ginseng's brain-boosting properties. Nevertheless, controlled studies have shown mixed results,[8] suggesting that it may increase memory performance, although this can be accompanied by a decrease in attention. Serious problems can occur from taking too much ginseng, since it has effects involving the adrenocortical hormones. It's best taken with care, or under the supervision of an experienced medical practitioner, and despite its popularity, the scientific jury is still out on this ancient supplement.

In Real Life

Education and information are the key to the safe and smooth running of your brain. Make sure you are fully informed about the things you put into your body and how they affect your thoughts and behavior, whether they come from multimillion-dollar drug companies or the local grocer. Also, don't overlook the everyday maintenance [Hack #69] of your body that can keep your mind optimally tuned.

End Notes

1. McCabe, S.E., J.R. Knight, C.J. Teter, and H. Wechsler. 2005. "Non-medical use of prescription stimulants among U.S. college students: prevalence and correlates from a national survey." *Addiction*, 100 (1): 96–106.

2. *http://www.washingtonpost.com/ac2/wp-dyn/A61282-2002Jun16? language=printer*.

3. Johnson, S.A., and V.F. Simmon. 2002. "Randomized, double-blind, placebo-controlled international clinical trial of the ampakine CX516 in elderly participants with mild cognitive impairment: a progress report." *Journal of Molecular Neuroscience*, 19 (1-2): 197–200.

4. Meikle, A., L.M. Riby, and B. Stollery. 2004. "The impact of glucose ingestion and gluco-regulatory control on cognitive performance: a comparison of younger and middle-aged adults." *Human Psychopharmacology*, 19 (8): 523–535.

5. Kennedy, D.O., A.B. Scholey, and K. Wesnes. 2000. "The dose-dependent cognitive effects of acute administration of ginkgo biloba to healthy young volunteers." *Psychopharmacology*, 151: 416–423.

6. Scholey, A. 2004. "Chewing gum and cognitive performance: a case of a functional food with function but no food?" *Appetite*, 43: 215–216.

7. Magill, R.A., W.F. Waters, G.A. Bray, J. Volaufova, S.R. Smith, H.R. Lieberman, N. McNevin, and D.H. Ryan. 2003. "Effects of tyrosine, phentermine, caffeine D-amphetamine, and placebo on cognitive and motor performance deficits during sleep deprivation." *Nutritional Neuroscience*, 6 (4): 237–246.

8. Vogler, B.K., M.H. Pittler, and E. Ernst. 1999. "The efficacy of ginseng. A systematic review of randomised clinical trials." *European Journal of Clinical Pharmacology*, 55 (8): 567–575.

—*Vaughan Bell*

HACK #74 Snap Yourself to Attention

Change your behavior by rewarding your hard work and punishing your slacking off.

As a self-aware animal, you can motivate yourself with the techniques of animal training. You can increase the number of your accomplishments by defining measurable units of achievement, breaking down your project into those units, and then giving yourself either a fixed reward per accomplishment or a fixed negative punishment for each lapse—or both, if you prefer.

First, a word about definitions. Behaviorist and animal trainer Karen Pryor has a concise, classic definition of a behavioral *reinforcer*:[1]

> A reinforcer is anything that, occurring in conjunction with an act, tends to increase the probability that the act will occur again.

Technically, punishment is not a reinforcer, since it *decreases* the likelihood an action will occur. (Of course, sometimes that's what you want.) We won't go into the intricacies of the behaviorist definitions of positive and negative reinforcement, which probably don't mean what you think they do.[2] There is even debate among behaviorists whether self-administered reinforcement, as in this hack, is a coherent concept or is better defined as something like "self-monitoring."[3] For the purposes of this hack, we'll just stick with the commonsense definitions of reward and punishment most people already have.

In Action

Break down your task into accomplishment units. If you're a *Getting Things Done*[4] fan, you can define these units as being the size of a *next action*, which is a single task of a few minutes or less that you can easily do to further a project.

When you accomplish one of these task subunits, reward yourself. Here are a few examples of rewards you can give yourself, paired with typical accomplishments you might want to reward:

- Surfing the Web for half an hour after writing five pages of your paper
- Having a piece of cake after studying 20 pages of your astronomy textbook

- Socializing for an hour after cleaning three rooms of your house

If you fail to accomplish one of the units of accomplishment, or exhibit some unwanted behavior, you can punish yourself, although you might find a reward-only system to be effective as well.

Here are a few examples of punishments you can give yourself, paired with typical unwanted behaviors:

- Burning a $20 bill for each day your paper is late
- Keeping all media (TV, radio, Internet, etc.) turned off all day whenever you break your budget
- Wearing a rubber band around your wrist and snapping it every time you bite your nails[5]

It's important to keep your reward or punishment immediate, consistent,[6] and proportionate. Snapping a rubber band every time you bite your nails is OK, but burning a $20 bill every time you do so is probably excessive.

Rewards (such as candy or a kiss) are usually preferable to punishment (such as an electric shock), not only because they're more ethical, but also because they're more effective.[7] However, since you'll be rewarding and punishing *yourself* in this hack, and you presumably know how much is excessive, you can use any combination that works for you.

You might need to experiment to hit the effective zone between excessive punishment or reward, and punishment or reward that's too weak to be motivational. A good place to start is to monitor your anticipatory feelings: do you look forward to your self-administered reward or cringe at a possible punishment? Do you feel motivated to get on with your task? If your punishment is excessive, you might just give up, so don't be too hard on yourself.

Excessiveness is only one of the problems that can occur if you're attempting to change someone else's behavior. If the subject isn't human, she might also misunderstand which behavior you're trying to change. However, when you're working with yourself, the problem of conveying the reason for the reward or punishment is lessened. As you eat your slice of cake, or hang out with your friends at the coffee house, reflect on your accomplishment to connect your behavior with the way you hacked it.

In Real Life

If you find the carrot and stick aren't working, maybe they're not big enough, or maybe you're not applying them consistently enough. Ask your friends to catch you in the act. They might really enjoy watching you snap yourself with a rubber band or burn a $20 bill. They might also have good ideas about what punishments and rewards you can use; it can be even more fun to solve this kind of problem for someone else.

As far as carrots go, my wife Marty and I are comedy junkies, so every time we finished writing and editing a hack for this book, we allowed ourselves a half-hour of video comedy. (For this reward, it helps to have a TiVo and access to some large video collections.)

Half an hour of comedy per finished hack proved highly motivating, but we made a basic mistake. We began lengthening the time we spent watching video and the frequency with which we did it, instead of decreasing our reward or keeping it the same. We started slacking off, and for a few days did little work on the book. It's not that we watched video all the time instead, but we took such long reward breaks that the reward became a distraction and interfered with our mental momentum **[Hack #65]**. When we cut back on how much we rewarded ourselves per unit of work, the amount of work we did increased.

As an example of the stick, I have several annoying speech habits, such as repeating myself and saying "um," "uh," and "like," so I've tried snapping my wrist with a rubber band when these habits crop up. Initially, I found that the rubber band made me acutely (painfully) aware of what I was saying and how I was saying it. My wife reports that now, every time I snap the rubber band, my bad speech habits stop for up to half an hour. Although I am now much more aware of my slip-ups, I've also asked her to point them out in case I don't catch them myself. Since she's the one who has to listen to the speech habits most, she's taken to helping me with a certain amount of glee. Perhaps I'll be able to completely eradicate these habits by snapping myself to attention.

End Notes

1. Pryor, Karen. 1999. *Don't Shoot the Dog! The New Art of Teaching and Training*, Revised Edition. Bantam Books.

2. Huitt, W., and J. Hummel. 1997. "An Introduction to Operant (Instrumental) Conditioning." Educational Psychology Interactive (Valdosta, GA: Valdosta State University). Retrieved October 18, 2005 from *http://chiron.valdosta.edu/whuitt/col/behsys/operant.html*.

3. Epstein, R. 1997. "Skinner As Self-Manager." *Journal of Applied Behavior Analysis*, 30: 545–569. Retrieved June 2, 2005 from *http://seab.envmed.rochester.edu/jaba/articles/1997/jaba-30-03-0545.pdf*. (This is a funny and eye-opening article on how pioneering behaviorist B.F. Skinner regularly changed his own behavior.)

4. Allen, David. 2003. *Getting Things Done: The Art of Stress-Free Productivity*. Penguin Books.

5. Norton, Ken. 2005. "Lifehack: how to stop biting your nails." *http://heynorton.typepad.com/blog/2005/06/lifehack_how_to.html*.

6. Ellis, Albert, Ph.D. 1988. *How to Stubbornly Refuse to Make Yourself Miserable About Anything—Yes, Anything!* Lyle Stuart. Despite the goofy title, this is an excellent book about becoming a more rational person. Chapter 16 has some especially good information about integrating behavioral therapy with Rational Emotive Behavior Therapy [Hack #57].

7. Epstein, R. 1997.

HACK #75 Assemble Your Mental Toolbox

Knowing many mental techniques is good, but having a few "tried and true" mind performance hacks always sharpened and ready to use can be extremely useful.

Despite my familiarity with a wide variety of mental techniques, I sometimes forget to use mind performance hacks when I need them most. For example, many times I could have used hypnosis [Hack #61] for self-motivation, but I simply didn't think of it in the moment. Similarly, in the middle of an argument, it's not easy to remember that the neutral point of view [Hack #64] can be a useful tool for settling disputes.

How can you remember to use mind performance hacks when you need them? You will need to do a little preparation in the form of outlining and memorization, but after that, you need to form only one habit: haul out your mental toolbox in moments of perplexity.

In Action

To create your personal mental toolbox, you'll need to follow five steps:

Inventory your mental tools
> The first step is to collect the hacks you find most useful. This book is a good place to start, of course, but many other books are available on becoming a better thinker, and you might have some hacks of your own. Use your catch [Hack #13] to collect your hack inventory as you think of it. These hacks will form the *tool* part of your toolbox.

Consider which tools work well together
> You might want to include a hack in your toolbox because it works well with another hack. For example, the two hacks beginning with *H* in the following section (Hypnosis and Heroes) work well together, as do the two *S* hacks (Seeding and SCAMPER).

Create a mnemonic "box" in which to put the tools
> You'll need to create a mnemonic to hold your hacks together. This is the *box* part of your toolbox. You might want to use an alphabetic mnemonic such as SCAMPER [Hack #22], or you might use a memory palace [Hack #3].

Tune your toolbox
> Never be afraid to tinker with your toolbox. J. Baldwin of the Whole Earth community had a portable machine shop in an unfolding walk-in van that he called the *Highly Evolved Toolbox* because it was slowly refined over 50 years.[1] Let your own toolbox evolve in a similar way, and leave some spare *slots*, or some other way you can edit your list.

Keep your tools sharp
> It should go without saying, but a tool in your toolbox that you can't use is just mental clutter. Learn to use the mind performance hacks on your list. Practice them from time to time if necessary. The same is true of tools that don't turn out to be as useful as you thought they might: if you put something in your toolbox and then find that you don't use it, take it out and try something else.

In Real Life

Here's a small sample mental toolbox. Different people have different needs, so they will have different toolboxes. Your own toolbox might look quite different from this list.

The hacks I selected for my toolbox had the initials *AFGHHNOPSS*. I took these letters and created the anagram *GH'S FAN SHOP*. Since *GH* in my personal Dominic System list [Hack #6] is represented by George Harrison, I expanded the mnemonic to *George Harrison'S FAN SHOP* and represented this mentally as a shop for fans of the late Beatle, with a mechanical George Harrison out front, playing guitar and tapping his foot in his characteristic way.

This imaginary storefront could become the beginning of a memory palace [Hack #3] (as described later in this section), but in my case, it serves only to remind me of the acronym, which stands for the following hacks:

G: Gott's Principle [Hack #45]
> Estimating the duration of any phenomenon with simple math.

H: Hypnosis [Hack #61]
> Motivating yourself with persuasive self-talk. You might find it useful to combine this hack with the other *H* hack, Heroes, since you can use hypnosis to enter deeper into the worldview of your chosen hero.

S: Seeding [Hack #19]

Picking a random object from your environment as the seed of a brain-storming session. You might find it useful to combine this hack with the other *S* hack, SCAMPER.

F: Forced connections [Hack #20]

Combining simple elements in random ways—or every possible way—to develop new and complex ideas.

A: Analogical thinking [Hack #25]

Finding what's similar about two patterns and extrapolating from one to the other.

N: Neutral point of view [Hack #64]

Defusing quarrels by rationally analyzing the discussion.

S: SCAMPER [Hack #22]

A toolbox within your toolbox that contains a set of brainstorming strategies. You can use SCAMPER to operate on the output of the other *S* hack, Seeding.

H: Heroes [Hack #31]

Keeping a compact set of alternate personalities within your toolbox (they fold up nicely). You might find it useful to combine this hack with the other *H* hack, Hypnosis.

O: Onar [Hack #28]

Plumbing your unconscious with a tool that's almost as simple as falling asleep.

P: Priming [Hack #12]

Priming the pump of your brain when long-term memory comes up dry.

Here are the first three items from a hypothetical memory palace [Hack #3] that you might use as an alternate way to memorize your toolbox, or to reinforce the alphabetic list. In keeping with the *GH'S FAN SHOP* mnemonic, I'll assume that the store is called George Harrison's Crackerbox Memory Palace and Hardware Store.

I walk past the storefront with the mechanical figure into the bright interior. Various tools are tucked into the nooks and crannies [Hack #4] of the shop:

Gott's Principle

In the near-left corner of the store is a pile of compasses; their arms form triangles, each of which contains a single Masonic eye (*Gott* means *God* in German). I pick one up and find that the arms of the compass will stretch to measure anything.

Hypnosis
> On the left wall is a display of rotary saws. Printed on each blade is a black-and-white spiral design. When I turn one on, the spiral rotates slowly and I find myself becoming sleepy, sleepy....

Seeding
> In the far-left corner is a barrel full of seeds with a big zinc scoop. When I look at a seed closely, thunder and lightning sound, and I see it's a tiny brain—a brainstorm seed!

Everyone carries around a set of mental tools that they habitually use to solve problems, collected over many years. This hack enables you to design your toolbox *consciously* and to always have your tools within easy reach.

End Notes

1. Baldwin, J. 1998. "The Ultimate Swiss Omni Knife." *Whole Earth Review*, Winter 1998.

See Also

- If you have your own "highly evolved" set of physical tools that you want to take everywhere, you might find that remembering 10 things to bring [Hack #1] will help. You might think of this hack as the mental-tool counterpart to that one.

Index

We'd like to hear your suggestions for improving our indexes. Send email to *index@oreilly.com*.

modulo arithmetic
  casting out elevens, 142
  casting out nines, 143
  casting out sevens, 165
  defined, 140
  overview, 165
momentum, mental, 72, 256–258, 292
monoamine oxidase inhibitors
       (MAOIs), 288
mood, mental exercises for, 65
Moravec, Hans, 96
morphological forced
       connections, 73–80
Morse code, 207–211
MotivAider timer, 63, 64, 65
multiplication
  aliquot parts, 134
  binary digits and, 151
  calculating digit sums, 140, 141, 142
  Chisenbop, 149
  Likert Scale and, 171
  mental math tricks, 129–132
  powers of 10, 133
multiplication tables, 128
Mumford, Alan, 55, 57
murse, 3
music
  creativity and, 104–107
  memory and songs, 33–37
  Oblique Strategies, 87
  Oulipo group on, 93
  Solresol and, 201
  thinking, 104–107
"must" statements, 232

N

naive mind, 252–255
nat language, 194, 195, 197
negative brainstorming, 169
negative mental momentum, 72, 256,
       258
neologisms, 196
neurotransmitters, 285, 287, 288
neutral point of view (NPOV),
       252–255, 293, 295
Newton, Isaac, 112
nibbana, 236
nines
  casting out, 140, 142, 143
  divisibility by, 138

nirvana, 236
nooks-and-crannies technique, 11–14,
       23
norepinephrine, 288
notes
  meditation and, 238
  stage fright and, 212
noun phrases, 220, 222
nouns, 17, 52, 53
NPOV (neutral point of view),
       252–255, 293, 295
number rhyme system
  mnemonic catch, 122
  overview, 2–4
  pegs and, 10, 14
number shape system
  five-digit numbers, 20
  mnemonic catch, 122
  overview, 4–6
  pegs and, 14
  remembering dreams, 111
numbers
  associating as shapes, 4–6
  associating with uses, 134
  digit sums, 140–144
  Dominic System and, 19–22
  doubling, 260
  friends of friends, 135, 136
  Morse code, 209
  numeracy and, 127
  rearranging for addition, 128
  recoding, 38
  rhyming words and, 2–4, 29–33
  special properties of, 131, 133–137
  (see also mental math)
number-to-consonant conversion, 30
number-to-letter conversion, 19
numeracy, 127
nutrition, mental fitness and, 272–276

O

object and location, associations in, 278
Oblique Strategies creativity deck, 72,
       87, 90, 93
O'Brien, Dominic, 10, 14, 18–22
Observation Deck, 89
Observer, 239
Okrand, Marc, 199
omega-3 fatty acids, 274
onar (oneiric sonar), 107–111, 197, 295

Colophon

The image on the cover of *Mind Performance Hacks* is a stopwatch, a time-piece that begins counting seconds at the push of a button and stops when the button is pushed again. Stopwatches are used to keep time when the measurement needs to be extremely precise, such as in races or scientific experiments. Although all stopwatches used to be mechanical, like the one on the cover, some now also come in the electronic variety, displaying the time they keep digitally on an LCD.

The cover image is from Stock Options Sports CD. The cover font is Adobe ITC Garamond. The text font is Linotype Birka; the heading font is Adobe Helvetica Neue Condensed; and the code font is LucasFont's TheSans Mono Condensed.

Better than e-books

Buy *Mind Performance Hacks* and access the digital edition FREE on Safari for 45 days.

Go to www.oreilly.com/go/safarienabled
and type in coupon code SQEP-G3TF-SWQ1-M1FE-13X7

Search
thousands of
top tech books

Download
whole chapters

Cut and Paste
code examples

Find
answers fast

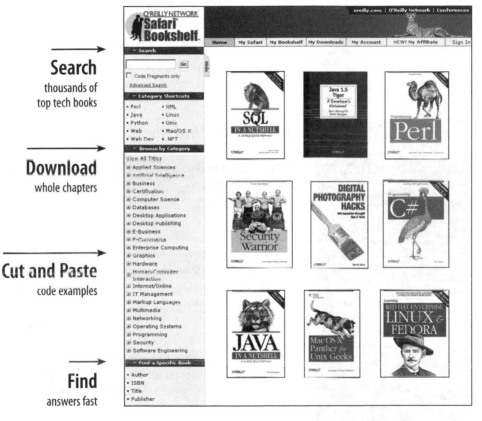

Search Safari! The premier electronic reference library for programmers and IT professionals.

Hack Your Head

Mind Hacks shows you
how your brain works.
Mind Performance Hacks
shows you how to
make it work better.

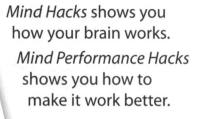

Mind Hacks
By Tom Stafford
& Matt Webb
November 2004
ISBN 0-596-00779-5
394 pages
$24.95 US, $36.95 CAN

Praise for *Mind Hacks*

"[*Mind Hacks*] makes a wonderful annotated bibliogra-
phy, with a light touch of hackish humor that inspires
further reading."

—*New Scientist*

"*Mind Hacks* is a rewarding mind trip, one that stretches
the boundaries of what we know about how the brain
works and playfully presents that information in an
engaging, thought-provoking way."

—*Blogcritics.org*

"If you've always wanted to get closer to your cerebel-
lum but never plucked up the courage to take that DIY
neurosurgery course, this is the book for you."

—*The Guardian*

"Mucking about in the wetware is a great way to spend
those long gray winter months…[W]ith the authors'
collection of tricks and tips for defragging, rebooting,
and fine-tuning your mind, you'll be in fine shape
come summertime. You may still look like crap in a
bathing suit, but you'll be smarter than ever and fully
in control of your own user interface."

—*Wired News*

O'REILLY®

Related Titles from O'Reilly

Hacks

Access Hacks

Amazon Hacks

Astronomy Hacks

Blackberry Hacks

BSD Hacks

Car PC Hacks

Digital Photography Hacks

Digital Video Hacks

eBay Hacks, *2nd Edition*

Excel Hacks

Flash Hacks

Firefox Hacks

Gaming Hacks

Google Hacks, *2nd Edition*

Google Map Hacks

Greasemonkey Hacks

Half-Life 2 Hacks

Halo 2 Hacks

Hardware Hacking Projects
for Geeks

Home Theater Hacks

iPod & iTunes Hacks

IRC Hacks

Knoppix Hacks

Life Hacks

Linux Desktop Hacks

Linux Multimedia Hacks

Linux Server Hacks

Mac OS X Panther Hacks

Mapping Hacks

Mind Hacks

Mind Performance Hacks

Network Security Hacks

Nokia Smartphone Hacks

Online Investing Hacks

Palm & Treo Hacks

PayPal Hacks

PDF Hacks

PC Hacks

PHP Hacks

Podcasting Hacks

PSP Hacks

Retro Gaming Hacks

Skype Hacks

Smart Home Hacks

Spidering Hacks

Swing Hacks

TiVo Hacks

Visual Studio Hacks

VoIP Hacks

Web Site Measurement Hacks

Windows Server Hacks

Windows XP Hacks,
2nd Edition

Wireless Hacks, *2nd Edition*

Word Hacks

XML Hacks

Yahoo! Hacks